SCIENCE FOR HANDICA

HUMAN HORIZONS SERIES

SCIENCE FOR HANDICAPPED CHILDREN

Alan V. Jones

A CONDOR BOOK
SOUVENIR PRESS (E & A) LTD

Copyright © 1983 Alan V. Jones

First published 1983 by Souvenir Press
(Educational & Academic) Ltd,
43 Great Russell Street, London WC1B 3PA
and simultaneously in Canada

All Rights Reserved. No part of this publication
may be reproduced, stored in a retrieval system,
or transmitted, in any form or by any means, electronic,
mechanical, photocopying, recording or otherwise without
the prior permission of the Copyright owner

ISBN 0 285 64974 4 casebound
ISBN 0 285 64969 8 paperback

Photoset and printed in Great Britain by
Photobooks (Bristol) Limited
Barton Manor, St Philips, Bristol

CONTENTS

Acknowledgements	8
Preface	9
PART ONE: SCIENCE AND THE HANDICAPPED CHILD	13
Introduction	14
Science within the School Curriculum	14
The Needs of the Children	18
Concepts and Topics	20
Various Methods and Approaches to Teaching Science	22
Numeracy and Literacy	25
Apparatus and Equipment	25
Aids to Learning and Communication	26
Planning a Science Curriculum	26
Evaluating Progress	33
Notes for Parents	33
PART TWO: SCIENCE EXPERIMENTS	37
Introduction	38
Chemicals around the Home	44
1 Colour Chemistry	46
Chromatography	47
Smarties and Spies	48
Don't be a Forger	49
Chromatography Pictures and Spots before your Eyes	50
More Pictures	51
Blotter Fish	52
The Beautiful Butterflies	53
Chromatography and Coloured Backgrounds	56
2 Acids and Alkalis	58
Colour Changes	60
Colour Changes with Red Cabbage	62
Reactions with Acids	63
Acid is the Opposite of Alkali	64

The Gas from Stomach Powders	66
Secret Messages	67
Changing Colours	68
Coloured Flowers	70
Chemical Pictures	71
Electricity	72
Static Electricity	76
Circuits – 1, 2, 3	78
Conductors and Insulators	82
Question and Answer Board	83
Electromagnets	86
Rob the Robot	87
Making and Using Magnets	89
Writing with Electricity	91
Some Problems	92
Symbols Used in Electricity	94
Ice and Water	95
What is Ice?	98
How Quickly does Ice Melt?	99
Can You Watch Ice Melt?	101
Lift a Lump of Ice with String or Cotton	102
What Happens when you put Salt on Ice?	103
Can You Believe your Ice?	104
Arctic Water to the Sahara Desert	106
Houses of Ice	108
To Find What Effect other Objects Have on Ice	109
Foams	111
Foams around the House	114
Does the Bathroom Sponge Contain Air?	115
Do All Foams Soak up Water?	117
Squashability of a Sponge	118
Boats	120
Divers	122
Submarines	124
Swinging Water	126
Candles, Light and Hot Air	127
Fire and Extinguishers	130
Candles and Flames – a few experiments	132
Candles Burning in Jars	137
Lights and Lamps	139

Bending Light around Corners 141
Writing with Mirrors 142
Hot Air 143
Temperatures Measured with Hot Air 145
Getting Dried Out 147
A Project 149
Strengths of Materials 150
Secret Strength 153
Bridges on Windy Days 155
Safe as Houses? 156
Straw Building 158
The Geodesic Dome 159
Can a Piece of Paper Hold up a Brick? 160
Make a Strong Building 162
Fibres, Strings and Ropes 164
Woven Fibres 166
Push, Pull Forces 168
Using Push, Pull Forces 170
Let's Make a Machine 172
Balance Machine for Weighing very Small Things 173
Some Things to Think About 176
Coloured Windows 179
TV Adverts 181
To Measure How Good Sticky Tape Is 182
XL 20 183
What is a Whammy-Diddle? 184
Happy Plants 186
A Mathematical Twist 188
A Few Problems 190

Appendix I 193
  Mainly for Teachers, Inspectors and Educators 193
Appendix II 205
  Basic Categories of Some Handicaps 205
Appendix III 208
  Some Useful Aids for Independence 208
  References 212
  Some Scientific Equipment Suppliers 213
  Suggested Further Reading 213

Index 215

ACKNOWLEDGEMENTS

The material presented in the following pages owes much to the expertise of the teachers listed and also to other teachers and pupils from the schools they teach in.
The major contributors are:
    A. V. Jones, Physical Science Department, Trent Polytechnic
    Susan Tomlinson, Fountaindale School, Mansfield.
Other contributors are:
    David Coleman, Aspley Wood School, Nottingham
    John Young, University of Nottingham Hospital School, Queens Medical Centre, Nottingham
    Tim Williams, Jesse Boot Junior School, Nottingham
    Gerald Leach, Hephaistos School, Reading
    Joy Cox, Keith Jones, Helen Mitchell, June White, Marion O'Connor, Margaret Freeman
    John Carter, Physical Sciences Department, Trent Polytechnic
    Nottinghamshire Science Inspectors and head teachers
    Graphics by Jennifer Warren
    Typing by Shirley Rackstraw, Jean Scott, Eve Stevens, Jean Spencer
    Photographs by Val Cliff, Philip Sharrock and others of the above named team.

Thanks are due to the children of all the schools listed and others who have tried the material with great expertise and enthusiasm.

The finances for equipment and other expenses of the research behind this project were supplemented by grants from British Petroleum, Leverhulme Educational Trust, Trent Polytechnic Research Committee and the Manpower Services Commission who were able to provide the final impetus.

PREFACE

Discovering things for themselves has always been one of the valuable ways that children, and people generally, acquire an understanding of the things around them. When a letter lands through your letter box I expect that you, like the majority of people, look at the writing and see if by this and the postmark you can *predict* who the letter is from. Sometimes it is all too obvious, as bills are usually heralded by the words 'Electricity Board', or 'British Telecom', on the outside of the envelope. In your home you must have had the experience of taking possession of, say, a new cooker or other piece of electrical equipment, and have had to read the instructions carefully to see how it works, gradually becoming an expert by process of experimentation. We are going to look at a number of experimental areas in which you have to do something and then think about it and, if you want to become an expert, go on to read around the subject as well.

This book is a series of science activities for people who want to 'expand their horizons'. The people for whom it is written are:

1. Young people, aged 11–16 years, who are physically disabled.
2. The teachers and parents of the young handicapped person.
3. Curriculum planners in special education, headmasters, advisers, inspectors and teacher trainers, all of whom may find it helpful.

It is probable that the ideas in this book would be appropriate for pupils (and parents and teachers of such pupils) who are orthopaedically handicapped and usually have the manipulative ability to feed themselves, or who can benefit from watching others doing experiments, so being able to think and see the principles behind them.

This is quite a wide category and can cover both young and old people who are in wheelchairs or bedbound, whether at home or in hospitals; those with normal mental ages and those who are retarded and have learning problems, including some people who are mentally handicapped. In trying these experiments it has been amazing how they have helped to motivate those who have often previously lacked any motivation at all.

One of the contributors to the experiments sections of this book has three principles behind his science teaching in a special school. It must be colourful, it must be noisy and it must move. If an experiment doesn't fall into one of these categories, he doesn't do it.

We realised, when compiling the book, that we had to be selective, and because there were always more ideas arising out of any one experiment, we have suggested extension experiments for those who want to go further with a particular topic.

A little more about the two main parts of the book

Part 1 This was designed for the teacher and interested parent, special education adviser and inspector who would often like to know how to incorporate some science into the curriculum of the handicapped child. It outlines the reasons for doing science and gives a suggested concept list and sequence diagrams for syllabuses. Advice is given for parents working with a child at home.

Notes on available materials in science, which, with a little modification, can be incorporated into the school or home curriculum, are included in Appendix I at the end of the book.

Part 2 Science Experiments. This is a series of topic areas containing a large number of science experimental activities for the children to do in the classroom or at home. They concentrate on the often neglected areas of the physical sciences, mainly because there is generally a greater teacher and parent knowledge of the biological sciences and of nature study activities. In addition, experiments in the physical sciences can be organised to be short, selfcontained activities using easily available apparatus, and these can be assembled for the pupils when and where they want them (at home or school, or in the hospital bed). Experiments in the physical sciences have often been neglected because they were thought to be too difficult and complicated. This is not so, as the experiments in this book will show.

This book is not an 'O' level or CSE book and is not intended to be a comprehensive collection of experiments for such a course. It offers a selection of activities which can be carried out with cheap and easily acquired apparatus found in the home or classroom.

The experiments could be linked to the sections of the syllabus outlined in Part 1, or could form part of a total curriculum for handicapped pupils called, say, 'Living in the Modern Environment'; the science experiments could form the 'Science for Living in the Modern Environment' section of the school course.

The experiments were developed to be interesting and motivating for the child who is often shut in at home. If other people use them in further ways then this is a bonus.

Part One

SCIENCE AND THE HANDICAPPED CHILD

'It is evident throughout this research that physically handicapped children can learn to understand science concepts and can develop higher levels of reasoning skills, afforded appropriate opportunity.'[1]

INTRODUCTION

A number of schools and hospital schools devoted to the education of handicapped children often have little or no science in their curriculum: indeed there seem to be very few such schools which have a curriculum including practical science. The explanations given by teachers run along these lines: practical science is a safety risk; there is no member of staff able or willing to teach science; there are no laboratory facilities; there is no money; there are difficulties covering the present curriculum in the time available; science is not appropriate for handicapped children.

Education without even a sprinkling of science is not a balanced education and ignores the fact that science can be a great motivator for all children. B. E. Thompson has noted that 'science can become an unsuspected ally in the struggle to provide success for the handicapped child who seldom enjoys success in school.'[2]

Although a number of individual teachers have started to tackle the problems of doing practical science with some handicapped pupils, their excellent ideas are not widely known; too often they are working in isolation.

Before planning a science curriculum, the following questions need to be considered: Which pupils can cope with practical science? What scheme is appropriate? How, and by what methods, can science be introduced? Does the science have to be examination-orientated or can satisfying experiments be used by non-examination classes?

The few available books on the subject, mainly from the USA, reveal that science for certain groups of handicapped children has been well developed; blind and hearing-impaired pupils seem to be particularly well catered for. Little, however, has been developed for orthopaedically-handicapped children or for the mentally handicapped child, and it is these neglected groups of children that this book has in mind.

SCIENCE WITHIN THE SCHOOL CURRICULUM

In comprehensive schools it is taken for granted that science for 11+ children is a basic necessity. In many primary schools the same assumption has been made, and the HMI reports of 1980 emphasise the increasing need for even more widespread use of science in the curriculum. Unfortunately the same assumptions have not been made for the special and hospital schools, and it is often left to the headteacher or individual teachers to encourage the inclusion of science into the curriculum of the handicapped child. The headteacher has the role of curriculum planner for the child and organiser of the professional strengths of the staff in the school. Perhaps the desire for science in the curriculum is present, but no member of staff is willing, able or available to try to introduce science into the classroom. Often, once the initial 'fear' of science is overcome and its great motivating force in the classroom has been realised, then the battle is won. This book cannot motivate the reluctant teacher or parent, but it might help to direct the thoughts of people contemplating what to do in science and how to do it.

Everyone, and we mean everyone, knows *some* science, and this book hopes to encourage the teaching of a balanced science curriculum, not just the things we know about and like. It deals with experiments or activities in simple physical sciences, as well as nature study and biology. The science all around us at this moment is more than just plants growing, but has to do with the ink on this paper, plastics, electricity, colour, weather, etc.

Because there are already many books available in the field of biology, catering for outdoor activities and nature study, the experiments provided in this book concentrate on the area of physical science which has not normally been considered with physically handicapped pupils in mind.

Which Children?

The ideas presented here are essentially for an activities-based science course designed mainly for children who fit the following criteria:

The child 1. must be capable of arm and hand movements similar to those required for feeding.
2. could be wheelchair bound, or able to sit at a table or sit up in bed, i.e. not necessarily totally mobile in the classroom or at home.
3. is not blind (although some blind children would benefit from doing some of the experiments).

4. has the mental capacity and reading age of eight years or above (or if lower, then works with someone able to explain the procedures).

Many children who have experienced extensive hospitalisation or home confinement may have lacked mental stimulation and could well have a developmental age below their chronological age. These activities could help in the maximising of their abilities.

The age of the child is not necessarily the criterion for deciding when to start some practical science experiments, but in general we have found that children of nine and above derive benefit from science activities of one sort or another. The upper mental age limit for this book would probably be 16, although it is acknowledged that many children might not reach this developmental level. Whatever a child's individual requirements, it is possible to select appropriate activities for him.

Children who are non-readers and have more restricted mobility and movements can still benefit if they work as part of a small group with children who can read or who are mobile.

What Type of Science?
Whenever possible science should involve doing experiments and activities and should not be restricted to learning facts and principles. The aims of a science course can be summarised as:

1. To provide safe, sound science, which is relevant and interesting.
2. To encourage the pupils to do experiments themselves and to do investigations suited to their own developmental level and disability.
3. To encourage a sense of achievement and a feeling of independence by working with scientific experiments.
4. To develop an individual approach to science, so enabling the teacher to spend time with each pupil or group.
5. To develop science activities and material on a topic basis, so that it can be incorporated into other classroom subject areas where possible.

On the subject of practical science activities, it has been encouraging to read the comments of Hadary who noted that 'when the appropriate adaptations have been made, the handicapped child is no more handicapped than the normal one.'[3]

The Need for Flexibility in a Science Scheme

The requirements of children in special and hospital schools and of those in comprehensive schools differ enormously, and therefore we have aimed at a flexible scheme which could fit the needs of many groups. A large amount of excellent science work has been developed during the last decade or more, to suit all ability ranges, and we have studied the published material in order to assess its suitability for handicapped children. We often found that apparatus, wording of worksheets and activities needed some modification before use with the handicapped child.

Science in the curriculum of the special or hospital school occupies a unique position in that it can be a link with the environment and the modern technological age. It can be the activity which motivates the child to talk, communicate, write, use number, draw graphs, paint, do art, etc. Science must not be seen as an isolated subject but as a series of topics and activities which overlap with other disciplines and subjects.

A group activity approach can sometimes help the slow learner who needs close, sympathetic support by the class teacher or parent. The activities must be carefully chosen to give each child a sense of involvement and help him or her to feel successful. These are important requirements, as often these children are unsuccessful in academic subjects and can feel a sense of rejection and isolation.

The construction of a total science curriculum is important, but each theme, topic or activity must be seen to be meaningful and interesting in its own right. The overall inter-connected topics woven into the context of a beautifully constructed syllabus cannot usually be seen by the slow learner; what matters to the pupil is the experiment in hand, while it is the teacher or parent who can see the importance of the whole syllabus.

THE NEEDS OF THE CHILDREN

It is difficult to describe the range of individual needs of each child, but we have attempted here to lay down some guidelines. The child needs to feel that he is doing worthwhile scientific activities. He must be educated to his own limits and not talked down to. He must feel a sense of achievement when doing an experiment. To feel a sense of independence is also important. The child's capabilities must not be underestimated.

The children in some schools are tested for IQ and other abilities, but for the activities included in this book the stage of educational development of the child is probably more appropriate. The influence of various theories of educational psychology has shaped most science curriculum plans; Piaget and commentaries by Shayer[4] in recent years have both emphasised the need clearly to define the stage of development of the child. A modified and more general plan useful for science education could be:

Early developmental stage (E_1)
The child at this stage can observe objects, apparatus and materials placed within a visual or touching distance and can play with them, but once these are out of his vicinity then the interest is lost. Only limited use is made of these objects and they are often dropped, knocked and sucked. These activities are usually observed in very young children, but may also be present, but not fully developed, in older children who are slow learners for one reason or another.

Early developmental stage (E_2)
The child at this stage begins to use objects and surroundings (including people) to achieve what he or she wants, e.g. builds bricks into a pattern, stands up objects which have fallen, begins to see the patterns in pictures, paintings, etc., and often tries to copy them.

Intermediate stage (I_1)
The child begins to organise objects and apparatus constructively and to achieve a set goal. He begins to see that there is a relation between cause and effect. He can see the relationship between, say, connecting a wire to a bulb and the light going on and begins to want to try other experiments. This stage could be accompanied by asking meaningful questions, such as

'Why?' or 'How?' although the explanation might be beyond the child.

Intermediate stage (I_2)
Here the child wants to discover the surroundings for himself and wants to do experiments constructively and systematically. He begins to understand the need for control when doing experiments.

Later stage (S_1)
Here the child is able to see the cause and effect relationship and can criticise poor situations, but is not reasonably able to construct better ones without help and advice. He or she begins to feel emotional involvement with experiments and often wants to succeed, so needs to know the correct answers to given problems. He is able to manipulate apparatus and equipment in an accurate and careful manner.

Later Stage (S_2)
Here the child is able to manipulate ideas rather than just apparatus, is able to think about difficult and abstract things and often makes suggestions and theories in difficult concept areas. He or she is able to construct logical problem-solving situations. Shayer showed that it is probable that only about 2% of the child population over 14 years are at this stage of development.

These artificial stages of development often overlap and a child can have quite a high ability in one subject area but be less confident and able in another.

The Slow Learning Child
This term must not be taken as a criticism of the child but just a convenient label to match activities to abilities. It is probable that the activities given in Part 2 could be approached and tackled by the majority of children at the I_1 and above stages but they will need close teacher or parent help and advice and probably help with reading and interpreting the instructions and diagrams.

It must not be felt that all the sections of the work task must be covered by all the children and probably, for some, only the early sections will be possible. This does not in any way detract from the motivation and enjoyment of the experiment.

The Good Thinker but Slow Manipulator
Some children can easily cope with the concepts given in the experiments, but find the manipulation of the apparatus difficult. In these circumstances either group work or close classroom or parent help is desirable. Usually the experiments assume the minimum manipulative skills that are needed to feed oneself.

Science experiments have been successfully done by many groups of children with IQs as low as 50 and also with children whose manipulative hand skills are very limited.

It is possible to design or choose scientific activities which could cater for most, if not all, of these groups of children and young people.

Some of the children who will use this book may suffer from the following disabilities:

Cerebral Palsy (Spastics)	Orthopaedic Handicaps
Spina Bifida	Visual Handicap
Muscular Dystrophy	Auditory handicap

A brief summary of the characteristics of children with these disorders is given in Appendix II, to help the reader to be more familiar with the possible problems.

The book may also be helpful to children in hospital as patients.

CONCEPTS AND TOPICS
The concepts listed opposite are those thought to be appropriate for a balanced, practical-based curriculum. They cover a wide range and are appropriate for various ages of children, the teaching style and approach being different for each ability group. Specific activities related to these concepts are given in Part 2 of the book.

The list of concepts is not presented in any teaching order but is given as a guide for those teachers seeking suitable areas of study which could be pursued with handicapped pupils (the topics are no different from those for able bodied children, but are listed here for the guidance of teachers wanting to write their own syllabuses, concept lists or scope and sequence charts).

Hot and cold
Temperature
Hot air
Keeping warm
Melting ice and snow
Warmth from the sun

Electricity and Magnets
Static electricity
Moving electricity, circuits
Changing forms of energy
Magnets
Electromagnets
Fuses and safety
Electronics, radio, etc.

Materials
Wood
Metals
Plastics
Fibres
Chemicals in the home
Minerals and stones

Time
Measurement of time
Speed, velocity
Acceleration
Earth's rotation
Astronomy

Living Things
Plants, seeds, leaves
Animals, large and small
Birds
Worms, snails
Ourselves
Common features
Hygiene (Home Economics)
Senses
Extension of senses and instruments

Light and Colour
Seeing
Shadows
Reflections
Refraction
Lenses
Rainbow
Photography

Structure and Forces
Structures involving paper, folding and tubes, bridges
Machines, crawlers, levers, wheels, balancing, pulleys
Machines in the home
Mass, weight, volume

Sounds
Vibrating things
Musical instruments
Hearing
Sounds on the move

Environment
Living things (see previous column)
Pollution, gases, liquids, solids, sounds, etc.

Air
Breathing
Air all around us
Water in the air
Moving air
Bases, oxygen, CO_2, acids

Water
Floating, sinking
Mass, volume
Dissolving, solvents
Washing, detergents

Skills and Processes involved in Science

These have been arranged in order of difficulty, with the earlier ones of a lower or easier priority. The later ones are more difficult and might not be reached by some pupils:

> Observing
> Measuring
> Practical and manipulative skills, making things
> Communicating (writing, diagrams, recordings, etc.)
> Ordering and classifying and looking for patterns in events
> Problem solving, cause and effect
> Projects (open-ended thinking) and designing apparatus and experiments.

VARIOUS METHODS AND APPROACHES TO TEACHING SCIENCE

The concepts and processes used in the experiments can be organised by the teacher, parent or curriculum planner in a number of different ways. Examples are shown below.

Science Interludes

This method introduces sciences using individual separate science interludes each week or day (based on a general progression of ideas). Examples could be separate experiments on:

> Question and answer board
> Some electrical circuits
> Colours and pictures
> Separation of colours
> Happy plants
> Writing backwards with mirrors

This approach is suitable for children whose attendance is spasmodic and who are often absent, or for those who are only on a short stay in hospital; it can also be used when short individual activities are needed.

In this method each separate interlude can be a self-contained activity, which can often be performed within the same lesson period. A different or related topic could be studied on the next occasion.

This approach is also suitable for home use when separate, short experiments are needed. (The experiments given in Part 2 lend themselves to this treatment.)

Problem Solving Activities
The pupil can be given a simple and specific problem to solve, perhaps with initial help from a structured work sheet. Experiments can be chosen as starters for these problems. The problems are related to the stage of development of the child. Often it is only the child at an advanced stage who can cope with this approach. Examples could be:

> Why does ice melt when salt is added to it?
> Why does a polystyrene ceiling tile feel warm but metal cold?
> How does an electric bell work?
> Is the black dye in felt-tip pens the same as black ink?

Direct Teaching or Circus Method
Some teachers prefer all the class members to be doing the same experiment at the same time. This could be a *more* expensive method of teaching any particular topic as one would need sufficient apparatus for everyone.

The more economical way would be to introduce a topic or area of science, say colour, and then distribute a series of different experiment boxes to do with colour to each pupil or group. Then each pupil or group can do their own different experiment and when it is completed go on to another experiment in the same general topic area. Thus over a certain period each child will have covered quite a wide area of activities in the topic area. This approach is also suitable in mixed ability classes, as each child or group could be given the most appropriate experiments without loss of prestige to the group. For example:

> Simple Electric Circuits
> (1) Can you light a torch bulb?
> (2) Can you light more than one bulb?
> (3) Can you change electricity into light?
> (4) Electromagnets

(5) Fuses
(6) Can you change electricity into heat?

Projects
These allow the pupils individually or as a group to work on a theme, or their work could be integrated to form an overall project. A project approach could also be adopted with home based pupils. Suitable experiments could be chosen to demonstrate a theme, together with random references to books, etc. For example:

1) Plastics all around us.
   One person could be researching how plastics are made. Another could make a few plastics. A further group or an individual could see which plastics are to be found around the home.

2) Make a clock or time-measuring device which accurately measures, say, two minute intervals.

Topic Work, Resource Based Learning, Project Work
In many schools at the junior age range or at the lower stages of the secondary school, a project or resource based learning approach is often used. This can cover many sub-areas of the school curriculum and can incorporate aspects of geography, history, English, maths, etc. The approach suggested in this book could be used or integrated into selected projects and could formulate appropriate science experiments for the handicapped child within these projects. When surveying the topic areas covered by Nottinghamshire schools, Michael Bassey[5] listed those most often used. If, for example, a topic is being considered on the local services, say the Police Force, then suitable science activities at the secondary or top junior level could be fingerprints (concept of touch), sound quiz for clues (concept of hearing), investigation of fibres (concept: materials and fibres), and forgery of cheques (chromatography). A further example could be the project entitled 'Things that change' and related concepts could be growth of skeleton; changes of energy (temperature, force); animal changes of colour and camouflage; reproduction in animals and plants; changes of temperature.

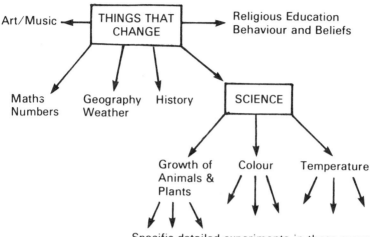

Specific detailed experiments in these areas are available in Part 2 of this book

NUMERACY AND LITERACY

Many handicaps have been shown to lead directly to a lack of spatial and mathematical skills and because of this the experiments should be selected with this in mind. (For this reason the experiments in this book have no mathematical bias.)

The literacy, reading ability and communication skills of a child are also, on occasions, not highly developed, and care has been taken when choosing activities that the words and format used are as motivating as possible.

APPARATUS AND EQUIPMENT

Whenever possible, a box of apparatus for each experiment would be a useful aid. The boxes would contain apparatus for complete experiments and their use would mean that the child had the minimum amount of unnecessary movement, looking for apparatus, etc.

The apparatus for a suitable science scheme need not be too technical in nature and could generally be put together by buying material and equipment from local shops and supermarkets. Some experiments for the later, secondary ages need more elaborate apparatus, but this is readily available from scientific suppliers and it is generally not too costly. Probably the most expensive piece of equipment is a microscope.

Knowing the financial limitations on all schools, we found that most of the science could be done with waste materials (yoghurt cups, wood, pieces of scrap metal, etc.) which are usually found around the home. This does not in any way degrade the science done, but it does need a storage space. More specialised apparatus can be obtained from the suppliers listed in Appendix III, and ideally should be stored in the drawers of a work table intended for the purpose – as a mini-laboratory.

AIDS TO LEARNING AND COMMUNICATION

In some cases the direct teaching method, coupled with a self-explanatory worksheet, would not be sufficient to emphasise the various skills and teaching points to the child. Experiments could be supplemented with synchrofax audio page worksheets or cassette tapes, explaining activities and experiments. This also helps those teachers and parents who initially lacked confidence about doing science. Often, after seeing the children's enthusiasm, everyone is convinced of its value.

The communication of the results or findings of an experiment can be recorded in any suitable way: by writing, filling in words, diagrams, drawings, pictures, audio tapes or simply verbally. It is essential to convince the pupil that scientists must communicate and record results; indeed, we hope that one outcome of doing science will be that children with communication problems will develop those skills, because of the motivating nature of the experiments and the desire to tell someone what they have found out.

PLANNING A SCIENCE CURRICULUM

The following pages will answer the Where, What, How and When of doing science, always assuming that you have already been convinced of the *WHY* of doing science.

Where do I Start?

The following scheme of work has been developed for children of mainly secondary age in a special school, with a wide range of developmental ages.

The general overall pattern of work could follow the following plan.

There could be two basic curriculum themes up to the age of sixteen:
1. Myself
2. Home and School Environment

The later stages of the curriculum could concentrate on Leisure and Hobby activities which could act as a good forerunner for the pupils' activities when they leave school.

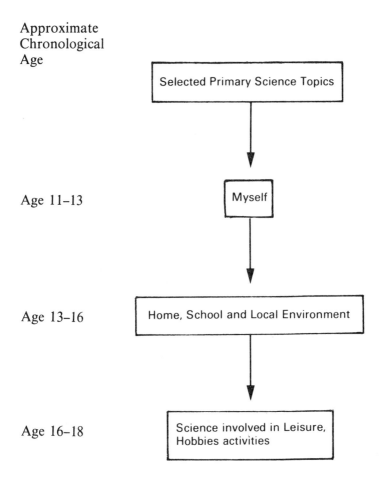

Network of Main Curriculum Areas, Topics and Concepts

(A more detailed breakdown of these areas follows the diagrams.)

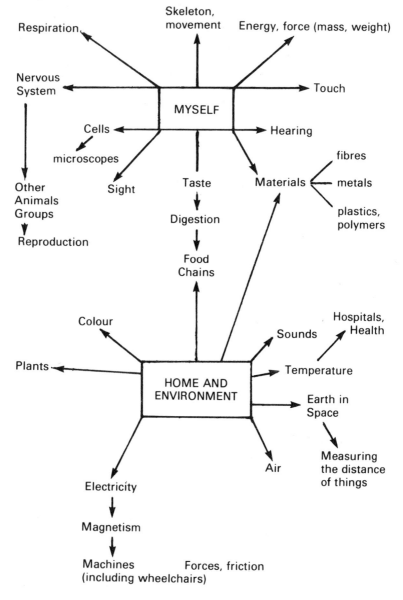

SCIENCE AND THE HANDICAPPED CHILD 29

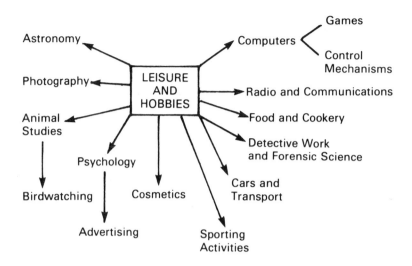

Breakdown of Content Within Each Sub-Area of the Curriculum
The important concepts given earlier are interwoven into the topics of the syllabus given below.

1 *11–13 Years: Myself, Senses and Extension of Senses*

Touch	Skin sensitivity – practical work: identification of objects by touch, fingerprints.
Sight	Observation of eyes, long and short sight, animal eyes, optical illusions, structure of the eye.
Hearing	Range of hearing, animal hearing, simple structure of an ear, direction of sound, ability to locate a range of sounds.
Taste and Smell	Testing sense of smell, taste discrimination, smell discrimination, taste buds, tricking your tongue, effect of smell on the taste of food, structure of the tongue.
Microscopes	Objects seen through hand lens, magnifying glass, microscope. Structure of a microsope, variety of microscopes, using and making slides of small objects, insects, plants and parts of plants.

Length	Concept of mm, cm, m. Measurement of objects, measuring distance travelled.
Weight	Various types of balances, gms and kgms, importance of weight, gravitational force, mass.
Classification of Animals	Molluscs – structure – marine and land. Worms – habitat, life cycle. Insects – metamorphosis – cabbage white butterfly. Reptiles – characteristics ⎫ Fish – characteristics ⎪ pupils study Birds – characteristics ⎬ specific example. Mammals – characteristics ⎭
Colour and Light	Why can we see objects? See in the dark? Which colours can be most easily seen? Shadows, reflections, lenses, mirrors, chromatography, periscopes, mixing colours, splitting of light.
Sounds	Vibrating objects cause sound, sound travels through air, making different notes.
Water	Floating and sinking, boats, flowing water, rivers, lakes, seas, surface tension, solutions and suspensions, dissolving.
Hot and Cold	Use of a thermometer, tricking our senses, movement of hot air, change of state, conduction, insulation, sun as a heat source.
Metals and Magnets	Force exerted on iron and steel, attraction and repulsion, compass, use of magnets.
Electricity	Electric circuits, electro-magnets, rubbed electricity, series and parallel connections, uses and sources of electricity.
Materials	Characteristics of familiar materials, natural and man-made fibres, reaction to heat, stress, water, uses and versatility of materials.
Plants	Structure and function of flowering plants – variety and habitat.

2   *13–16 Years*

Food Chains, Food Webs	An ecological habitat – study of a woodland, pond. The effect of pollution on a river. Chemical farming.

SCIENCE AND THE HANDICAPPED CHILD 31

Man	Digestion Respiration Skeleton and Movement Nervous system Circulation Skin Reproduction	} Structure and function
Cellular Structure	Plants – xylem – phloem – stoma – typical	} Structure and function
	Animal – typical – nerve cells – muscle cells – blood cells	} Structure and function
	Concept of cells as building bricks, microstructure.	
Levers, Force, Friction	Forces of gravity – pendulums, etc., balancing objects, stability, pushing, pulling, movement of objects along a curved path, levers, pulleys, gearing, bouncing, paper engineering.	
Electricity	Electric motors, volts, etc. Applications of electricity, making a bell, torches, question and answer card, scoreboards, electricity in the home.	
Water	The water cycle, evaporation, condensation, importance of water to life.	
Air	Necessary for burning and breathing. Moving air as an energy source, properties and composition of air.	
Earth in Space	Rotation of the earth, seasons, moon, planets in relation to the earth, stellar concepts.	
Plants	Photosynthesis, respiration, life cycle, classification of plants, bacteria and moulds.	
Materials	Chemical and natural materials – a comparison of their properties and uses.	

3 *16–18 Years*
Astronomy   Telescopes, moon maps, space travel.
Computers   Computer language, principles behind a com-

	puter, making a simple computer, simple programming, use of a computer, computers in industry, home.
Radio Communication	CB radio, language and communication, making of simple radios.
The Motor Car	Oil, 4 stroke cycles, batteries, electrical systems, hydraulic systems, cooling systems, corrosion, safety, friction, transport.
Psychology	Senses, sensory illusion, perception, reflexes, conditioning, learning, trends, opinions, attitudes, advertising, personality and personal relationships.
Photography	A pin hole camera, simple cameras, use of 35mm cameras, SLR cameras, developing film, enlarging, uses of photography. Photography without a darkroom.
Cosmetics	Skin, hair, teeth and nails. Making cosmetics, emulsions and skin creams, soap and toothpaste. Making up eyes, hair creams and essences.
Ornithology and Animal Study	Bird names, habits and habitats, wildlife in Britain. Ecology, technology and conservation. Animal care.
Food	Cookery and home economics.
Sporting Activities	Swimming, snooker, etc. (Selected appropriate activities)

Matching a Syllabus with Activities
Some possible science schemes already published, that match the topics given here, are suggested in Appendix I.

The experiments given in Part 2 of this book are not chosen to be an exact replication of all the topics listed in the syllabus given above, but have been chosen as enrichment experiments for handicapped pupils, to supplement school and home science topics and to demonstrate some of the concepts listed on the previous pages.

The experiments have been devised to exemplify that an activity approach to science is possible for handicapped pupils. They can be used to replace inappropriate experiments in some conventional science textbooks or science schemes.

How do I Know the Experiment will Work?

All the experiments listed in this book have been tested extensively with children, both handicapped and able bodied. Similarly, the experiments in other recommended schemes have been tried with many children and are known to be of use. It must be remembered that sometimes the initial skills of the experimenter are at fault and some experiments need to be done once or twice to practise the necessary skills. A research student sometimes repeats his experiments many, many times before he is satisfied with his skills and results.

Replacement Experiments

The lists and details of experiments in this book are not exhaustive, and if you see others or can think of better experiments, then add them to your curriculum.

EVALUATING PROGRESS

How do I Know I've been Successful?

When a child has finished a particular experiment and he or she has communicated the results in some way to the teacher or parent, it is useful to keep a progress chart. A suggested grid is shown overleaf.

You will be able to see from the categories listed at the tops of the columns, and from the individual record for each pupil, whether the child is making progress and which activities could be chosen for further mental stimulation.

You should also be able to see if the child is progressing in understanding and communication skills, etc.

NOTES FOR PARENTS

If you are going to do experiments at home then it is necessary, whenever possible, to let the child cope alone. Don't be always looking over his shoulder but allow a feeling of independence to develop. Encourage this by whatever means possible. None of the experiments are dangerous, and so the child can cope with any minor mishaps, as he has to do in his normal living.

Avoid forcing your child to do any particular experiment; each must be interesting and motivating, so that he will want to do it and, when it is completed, be motivated enough to do

Name.................... Year......... Term......... Age.........

Activity Names and No.	Date	Did the child show any of the following characteristics, either indicate ✓ or × or use an A-D (4 point scale)									Indicate what type of communication used and quality of work. Written, drawing verbal, etc.
		Observational skills	Raising Questions	Understanding of Concepts	Curiosity Exploration	Enjoyment	Originality	See Cause/Effect	Problem Solving	Cooperation with Others	

another experiment (but not necessarily straight away).

Motivation totally stimulated by reading a text or looking at a picture is difficult, and so it is generally assumed either that the child is interested enough to start an experiment or that you, the parent, have been interested enough to buy this book and can act as the stimulator of interest in the first instance.

Often one experiment leads to another, and if further totally open-ended 'experimenting' is done, then great care must be taken to ensure personal safety.

We have found that simple home experiments can often lead to greater communication skills (verbal, written work, pictures). Encourage the child to keep a record of his or her experiments in case a discovery is made, or in case someone else might like to see the results.

Many scientists started their interest in science by doing home experiments like these.

Further details about the educational development of handicapped children, and the help available to parents and teachers in the form of publications and courses, can be obtained from your local special education inspector (at the local education offices).

Part Two

SCIENCE EXPERIMENTS

The following pages include details of experiments which can be done at home, in any school classroom (no specialist laboratories are needed) or in hospital. The majority of the experiments use everyday materials and apparatus, so they have the advantage of being cheap and easy to perform.

The experiments have generally been planned as enrichment experiments for the school curriculum (including topic work, projects, etc.) and also to be used as motivational and interest activities for the child and young person.

INTRODUCTION

To the Experimenter at Home
For anyone who is confined to the house for any length of time, the immediate surroundings can become rather uninteresting. The kitchen equipment, living accommodation and the whole house become all too familiar. But suppose that even the familiar hides a new dimension, a new area of exploration! We know already that the kitchen is a place of experimentation; sometimes the cooking experiments work and occasionally they are failures. The kitchen is like a laboratory.

The equipment we use in the home, whether electrical (electric lights, radio, TV) or mechanical (the taps, hand food mixer, the knives, door handles) have all been designed after a series of closely observed experiments.

SCIENCE EXPERIMENTS: INTRODUCTION 39

Your home is going to become a laboratory, a design centre, and you are going to be the experimental research scientist. There are experiments for you to do and some of these have extension experiments or problems for you to solve. This may mean doing further experiments, designing your own experiments or on some occasions reading and searching for information in books and the local library. Sometimes you will have to just stop and think.

Remember, the experiments that you are going to do are unique; no one has ever done that particular experiment in that exact way before. Some people will have done similar ones, but never identical ones. You are truly a research worker. Because of that it would be a good thing for you to keep a notebook or file of what you observe and do. Keep a record of results and say how you could improve the experiment also.

This book of experiments is only a starter, and you might want to go on and do other experiments in other books, or to investigate further a particularly interesting topic.

The experiments in this book are all safe if you follow the instructions and all of them have been tested. They are designed for you to use your abilities and not emphasise your disabilities.

This section of the book is *not* an 'O' level or CSE text book; nor is it a book to be read from cover to cover. It is a book to help you *do* a scientific activity, investigation or experiment.

Don't be put off by the fact that you might not have done much science before. The instructions given are clearly written and easily understood.

There is no definite order to the experiments, and you can try any one you like and in any order. We have grouped them according to similarity of topics, eg a series of experiments with ice and water and another series using chemicals around the home, etc.

You can do one experiment one day and another one on another occasion. If you go into hospital, then take the book with you. There are some hospital schools where they do science experiments.

The experiments can be done individually or used as a group experiment with a small group of friends.

Some Notes before You Begin

This collection of experiments is designed for people who would like to do some scientific investigations in the home or school.

1. They are designed for all handicapped young people who have the ability and mobility to hand feed themselves. We have mainly kept in mind the orthopaedically handicapped person (a blind person can do a number of the experiments depending upon the degree of vision).

2. The experiments can be done by anyone (of any age), but some basic level of reading and ability to carry out instructions is assumed. It would be possible for a second person to read out the instructions and interpret them to the experimenter and still get a lot of scientific enjoyment out of the experiments.

3. The laboratory is right beside you – the living room and kitchen. Both are ideal places for scientific investigations, and the 'apparatus' we are going to use can easily be found around the home or purchased in a local shop.

   At the start of each experiment there is a list of the things you need to carry it out.

4. All the experiments are as safe as we can make them and there is no chance of you being blown up doing the chemistry experiments or electrocuted using the $4\frac{1}{2}$ volt batteries. Any safety precautions needed are given in the details of individual experiments.

5. Often, the first time you do an experiment you are just getting the feel of the apparatus, and sometimes you might have to repeat it a second time to give satisfactory results. All scientists have to do this.

6. Many scientists started their science by doing home experiments.

7. We have written this book so that you can take an experiment from any part of the book, but it is probably best to work through one section at a time.

8. Often one experiment leads to another, and if you do your own experiments, not in this or other books, please make sure they have been checked for safety.

IMPORTANT

Never use mains electricity for experiments.

Keep water and wet hands away from mains electrical plugs and lights.

Be careful when handling hot or boiling water or lighted candles, etc.

Take care using sharp instruments, knives, scissors, pins and needles.

Enjoy your experimenting. Science has given a lot of people a great deal of enjoyment, interest and knowledge of the things that are around us.

About the Experiments – for Parents and Teachers

For Teachers organising Science in the Home
The visiting teacher might like to use the science concept maps given in Part 1 as guidelines to what can be done with handicapped children, and then select specific science activities from the following pages.

For Parents
If, as a parent, you are looking for a single motivating activity, then it is possible to select individual separate experiments from this book, but if the child is going to spend a prolonged period at home, perhaps it would be advisable to take a more long term structured approach to the topics covered. For example, the theme of 'Chemicals around the Home' could be used together with the individual experiments on this theme given in the book. Alternatively, one might focus on a theme or project arising from an unscientific area, like 'Paper'. This project could be researched from books, etc, and relevant experiments could be selected to demonstrate aspects of this theme.

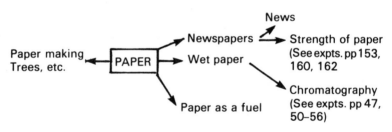

The experiments chosen for inclusion in this book have been selected to give all children a chance of doing some science for themselves. So that experiments can be matched to the children's abilities, they have been given a 1–4 starred rating.

* *   These experiments are generally simpler and can be attempted by children at the I_1 stage (see page 18).
* **   These need a little more understanding and more advanced manipulative skills, mainly for children at I_2 stage of development (see page 19).

*** These require some more advanced thought and manipulative skills and are suitable for children at the S_1 stage (see page 19).

**** These are quite advanced in the manipulation of ideas, apparatus and thinking. Often, these activities are sufficiently open ended to give the thinking child a mental challenge. (S_2 stage, see page 19).

A good way to select suitable experiments for the child would be to give him or her a single star activity as an introduction to a topic and then, if successful, quickly move to, say two or three star activities. Usually the extension to an experiment is at least the same star ranking, and often one star higher.

It has been generally assumed that the experimental details will be read or interpreted for the child by the parent, unless the child has the reading ability to cope with it himself.

The use of flames has generally been avoided in all but one section of the experiments (and this can be dispensed with if necessary). Also, we have avoided wherever possible those materials that could be dangerous (eg strong bleaches, strong acids, sharp knives, etc).

Some of the experiments might lead the child to want to follow a particular line of study at the local library or via other books.

CHEMICALS AROUND THE HOME

There are many useful and interesting chemicals around the home. Most of them are safe when used correctly.

The experiments in this chapter use acids and alkalis present in the house, together with coloured food dyes.

The first few experiments investigate chemicals using paper chromatography. The name chromatography is derived from the name for 'colour' – chrome – and 'on paper' – graphy.

When a coloured substance in solution is placed on a piece of absorbent paper (like blotting paper), then the colour soaks through at a characteristic rate, unique to itself. If we take a mixture of a few different substances, each substance will soak through at its own characteristic rate. The mixture can then be separated into its constituents. Our coloured mixtures will be felt tip pens and Smarties.

CHEMICALS AROUND THE HOME

EXPERIMENTS IN THIS CHAPTER

Colour Chemistry
Chromatography*
Smarties and Spies*
Don't be a Forger***
Chromatography Pictures and Spots before your Eyes*
More Pictures **
Blotter Fish**
The Beautiful Butterflies**
Using Chromatography to Make Colourful Backgrounds*

Acids and Alkalis
Colour Changes***
Colour Changes with Red Cabbage**
Reactions with Acids**
Acid is the Opposite of Alkali**
The Gas from Stomach Powders***
Secret Messages**(* *with help*)
Changing Colours**
Coloured Flowers***
Chemical Pictures***

Star Rating Reminder
*   Experiments that are relatively simple to do (if necessary, with help from parent or teacher).
**   Experiments requiring a little thought and manipulative skills.
***   Experiments involving thought and reasoning powers and manipulative skills.
**** Experiments, problems and questions requiring a lot of thought and reasoning and often the designing of experiments, etc.

COLOUR CHEMISTRY

Materials You will Need
Water soluble felt tip pens (a collection of colours)
Blotting paper (white)
Cotton wool
Tube of Smarties
Dylon cold water dyes (a selection of colours)
Container of water (ie jug)
Plate or saucer
Cup or jam jar, or margarine tub
Scissors
Wire coat hanger
Large bowl or aquarium tank

Preparation by Teacher or Parent
Each experiment is self-explanatory, but some preparation is needed for the experiment headed 'Don't be a forger'. You will need to make the 'forgery' of the cheque as shown and explained in the experiment.

Because the dylon dyes are highly coloured, surfaces which could be affected by spillage could be covered with newspaper or polythene. The experiment using the dyes need use only a very small amount of powder. Too much spoils the effect.

CHEMICALS AROUND THE HOME   47

Chromatography   *

You will need: A few water soluble felt tip pens
                Blotting paper
                Container of water (ie a jug)
                A piece of cotton wool
                Plate or saucer

What to do
Cut a piece of blotting paper so that it is just about the size of a plate.

Put dots of the black felt tip pen in a circle about the size of a 10p coin at the centre of the paper.

Now wet a small piece of cotton wool and squeeze out the excess water and put the moist cotton wool at the centre of the circle of dots.

Watch as the moisture gradually soaks outwards.

What happens to the dots?

What does this tell you about the colours used in making that particular black felt tip pen?

Try other makes of black felt tip pens.

Extension   **

Repeat the experiment, but this time use a selection of colours. Put dots of a number of different colours on your circle. Do not put them too close together.

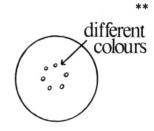

48 SCIENCE FOR HANDICAPPED CHILDREN

Smarties and Spies *

Suppose you were an industrial spy and you wanted to make a new range of coloured sweets and you wanted to find out what colours were in Smarties, how could you do it? Using chromatography, can you see how?

You will need: Some blotting paper
Tube of Smarties
Cup, jam jar or glass of water

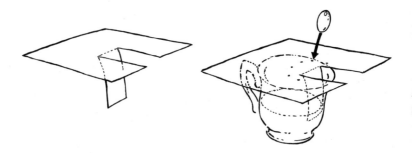

What to do
Take a piece of blotting paper, cut or rip a tongue in it like the drawing and fold the tongue downwards.

Lick the surface of the Smarties and put a spot of each colour in a ring around the paper tongue. Try to keep the spots as small and as highly coloured as possible.

Dip the tongue of the paper in the water and watch what happens to the colours of the Smarties.

Is the brown Smartie made of a brown dye?

Extension ***
How could you see if the dye (say an orange colour) used in Smarties is also present in another orange coloured sweet?

If they are the same orange dye they should move the same distance during the same time on the same type of paper using the same liquid to soak along it.

Try the same type of experiment with felt tip pens.

CHEMICALS AROUND THE HOME 49

Don't be a Forger ***

Problems
Use your knowledge gained from earlier experiments, to solve this problem!
Write a cheque on a piece of blotting paper like the one written below, using a *biro pen*.

Now, to forge this cheque, use a water soluble felt tip pen as shown below:
1 Add the words 'One hundred and . . .' in front of 'one pound'.
2 Add the number '10 . . .' in front of the 1.00 in the box.

Suppose you were a detective, how could you show that the above bank cheque was a forgery?

Hints
Remember that felt tip pens on blotting paper can be separated out using water and chromatography.

Chromatography Pictures and Spots before Your Eyes *

You will need: Blotting paper
A few small containers of cold water Dylon dyes
Container of water
Cotton wool

What to do

Carefully prick a hole in the container of the Dylon dyes and sprinkle a *very small* amount of each dye on to a square of blotting paper.

You don't need much powdered dye on your paper.

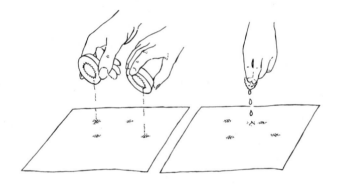

Now smooth out the dyes over the whole surface of the paper. You can do this with a dry paper hanky.

Take some cotton wool and dip it into some water, then drip the water from the cotton wool onto the sheet of paper with the dye on it.

You can see the coloured effect immediately and perhaps you could design a picture of, say, bunches of flowers in a vase or garden.

The pattern you make with this technique has never been put together in exactly the same way before. So your picture is unique.

Is it a masterpiece?

More Pictures **

You will need: Blotting paper
Felt tip pens
Cotton wool
Container of water

What to do
Use your felt tip pens to draw some shapes on your paper. You could draw

 or or

You could go over each shape with a few colours, one on top of another.

Now take a few pieces of cotton wool, moisten them and squeeze out the excess water. Now put the moist cotton wool at the centre of each shape you have drawn.

Leave the water to spread out until the picture is satisfactory.

Good effects are obtained if you let the colours merge one on another.

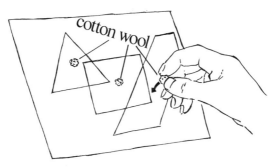

Extension **
You don't have to use blotting paper. You can use other materials also, for example a clean linen hanky (ask your Mum first). This type of experiment doesn't work very well with paper tissues or kitchen roll.

Blotter Fish **

You will need: Felt-tip pens
White blotting paper (20 cms × 10 cms approx)
Scissors
Small rectangular margarine tubs

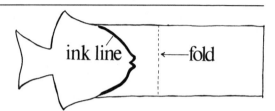

What to do

Draw a simple fish outline on your blotting paper in pencil and cut it out as shown.

Remember to leave an extra piece of blank paper in front of its mouth, and fold this over.

Carefully draw around its head with a black or brown felt-tip pen (or repeat the line with a number of different coloured pens).

Hook the folded paper over the rim of a margarine tub half full of water. Watch what happens.

When the fish is fully coloured, take it out of the water and lay it out to dry. Cut off the flap around the mouth and draw in a spot for an eye.

Now try this

Use another fish shape with the flap at the tail end. Try using different coloured pens.

Perhaps you can hang these fishes up in your room on black cotton or put them on the mirror with Blu-tak.

The Beautiful Butterflies **

Attractive symmetrical shapes can be made using chromatography. Butterflies are good to try because they have highly coloured wings.

You will need: 2 margarine tubs, one of them with a hole in the bottom
Scissors
Blotting paper (20 cms × 10 cms)
Felt-tip pens

What to do
Fold the blotting paper in half and draw on one side the shape of a butterfly. Cut it out and you will make a symmetrical butterfly shape. Draw in the outline of the body with a black felt-tip pen (as shown).

Refold the shape and push the body part through the slit in the margarine tub until it *just* dips into the water in the lower tub. The water should be only *just* touching the paper (this is important).

Watch while the colours spread out onto the wings. Remove the butterfly when the colours reach the wing tips. Lay them flat to dry, then draw on the eye and body markings.

Now try this
Draw some shapes of butterflies and cut holes in the wings before you start. Use some different symmetrical shapes such as flowers, or you could try drawing bird shapes.

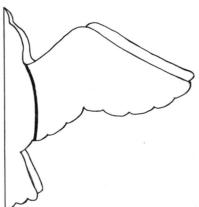

CHEMICALS AROUND THE HOME 55

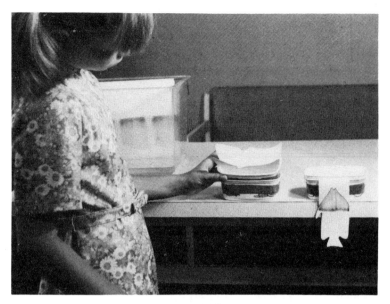

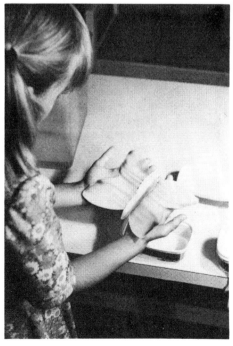

Using Chromatography to Make Colourful Backgrounds   *

This activity involves making unusual backgrounds onto which you can draw or stick a suitable picture. If you use a large sheet of blotting paper you can make two at the same time.

You will need: Blotting paper
Felt-tip pens
Wire coat hanger
Aquarium or large container

What to do
Fold the blotting paper in half. Each half can make a different scene. Draw a wavy or broken line with your coloured pens across both ends of the paper about 5 cms from the edge.

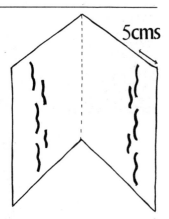

Put water in an aquarium or bowl, or in the sink, to a depth of 3 cms. Use the wire coat hanger (either in its usual shape or straighten it out) or a stick to make a rod which can rest across the top of the tank. Hang the double sheet on the rod allowing the ends to just dip in the water. The water must not touch the ink lines.

Watch the colours move upwards and remove the paper for drying when they are near the top. Dry the paper. Cut the sheet along the fold.

You could use black paper to make some silhouettes and stick them on the background (like a space ship or boat, birds, etc.)

Now try this
Try drawing wavy and broken lines at different heights to make even more stunning effects.

CHEMICALS AROUND THE HOME 57

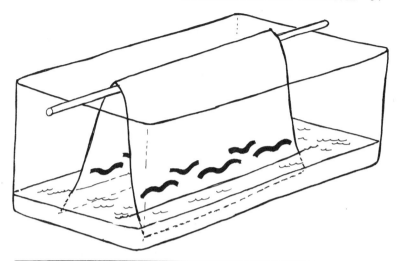

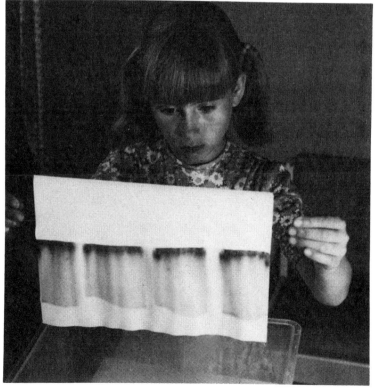

ACIDS AND ALKALIS

What is an acid?
These are either liquids or solids which have a sharp taste, and are able to give carbon dioxide from a carbonate or bicarbonate. They oppose or are the opposite of alkalis.

Some acids are weak, like vinegar (acetic acid), lemon juice (citric acid) and tartaric acid, and these are used in cooking and for flavouring foods.

Much stronger acids are used in car batteries (sulphuric acid), or for making explosive (nitric acid), and these acids are certainly not used as food additives. They would burn you severely. The sulphuric acid in car batteries will severely burn the skin and dissolve metals.

What is an alkali?
These are liquids which are the opposite of acids. When an alkali is added to an acid it cancels out the acid and makes a neutral solution.

Weak alkalis used in foods, etc., include bicarbonate of soda.

Stronger alkalis which are too strong to be used for foods, or which burn the skin, are chemicals like ammonia and sodium hydroxide. These are used for cleaning very resistant surfaces and are included in strong bleaches, oven cleaners, etc.

Some coloured chemicals change colour depending upon whether they are in acid or alkali solution. These substances are called 'indicators'.

The colour changes of chemicals, depending upon their environment, are not restricted to acid or alkali solutions. Iodine causes starch solution to change to a dark blue-black compound.

CHEMICALS AROUND THE HOME

Materials You will Need
Smarties
Water soluble felt tip pens
Bicarbonate of soda
Vinegar
Ammonia solution
Red cabbage or pickled cabbage
Moth balls
Coloured inks, or Dylon dyes
Table salt
Birthday candles
Plasticine
Alka Seltzer tablet
Bottle of iodine (from medicine cupboard)
Saucer
Tall jars or glasses (taller the better)
Spoon
Tumbler/glass or jam jar
Pen with nib
Plain writing paper
Water from cooking potatoes or rice
Paint brush
Cotton wool

Optional
Cobalt chloride ⎫
Copper sulphate ⎬ These can be bought at the local chemist shop, possibly.
Ammonia ⎭

60 SCIENCE FOR HANDICAPPED CHILDREN

Colour Changes ***

You will need: A tube of Smarties
              Felt tip pens
              Bicarbonate of soda (solid)
              Vinegar
              Ammonia solution (diluted with water)

CAUTION
Do not eat the Smarties after you have been experimenting with them.

What to do
We are going to see if the colours of some sweets and felt tip pens change when put in different solutions of common chemicals.

Take three saucers and put in each a little of the following household chemicals:

Take a few different coloured Smarties and dip one of each colour *briefly* in the separate solutions and then leave the Smartie to drain on the side of the saucer, out of the liquid.

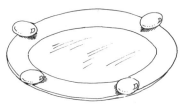

CHEMICALS AROUND THE HOME 61

The vinegar is an acid called acetic acid.
Bicarbonate of soda solution is a weak alkaline solution.
Ammonia solution is a strong alkaline solution.

The chemically opposite thing to an acid is an alkali.
Did any of the Smarties change colour when added to acid or alkaline solutions?

Do you think the colours would change back if you dipped the Smartie which had changed in the acid solution back into an alkaline solution? Try and see.

Extension   ***
Try the same type of thing by drawing lines on paper of various felt tip pens and try dipping them in the acid and alkaline solutions.

Make a diagram of your results.

Colour of original	Colour in vinegar	Colour in bicarbonate of soda	Colour in ammonia

Would any of these colours be any use to test to see if any other chemical around the home was an acid or alkali?

We call chemicals which change colours in acids and alkalis 'indicators'.

Take the best material which changed colour and test the following things:
Washing up liquid
Lemon juice
Orange squash
Soap and any other things you would like to try.

Colour Changes with Red Cabbage ∗∗

You will need: Some red cabbage (or pickled cabbage)
Vinegar
Ammonia solution
Some glasses or jam jars

What to do
You are going to put on a magic display. Take some red cabbage and boil it in water for about 20 minutes. The coloured water will be used for our experiments.

Into a jam jar or tumbler put about $\frac{1}{4}$ of the red cabbage water. It looks like red wine.

Add a few drops of ammonia solution until the colour just changes. You will have to stir with a spoon also.

red cabbage water

Add a few more drops and further colour changes occur.

You should get a purple to red to bluey purple to green (if you add a lot of ammonia).

The magic is to tell someone that you can change red wine into green Chartreuse! **But don't try drinking it!**

Extension ∗∗∗
You can try extracts from other plants also, say, the leaves or flowers of bluebells or beetroot or wallflowers.

The colour from plants can be dissolved out with meths also. The plant can be crushed with a small amount of meths and the coloured liquid can be used as above.

Reactions with Acids   **

You will need: Some *tall* jars (taller the better)
　　　　　　　Moth balls
　　　　　　　Vinegar
　　　　　　　Some coloured inks or food dyes
　　　　　　　Bicarbonate of soda
　　　　　　　Table salt
　　　　　　　A large jug

What to do
In a jug of tap water dissolve, a spoonful at a time, enough table salt so that a moth ball (of naphthalene) can *just* float on the surface of the water.

Now carefully and slowly add tap water, while stirring, until the ball *just* sinks.

Put some coloured ink or food dye in the water, just to make it pretty.

Now stir into the jug of water a teaspoonful of bicarbonate of soda, stir until it dissolves.

Now put the coloured solution in the tall jars.

Put a few moth balls in the water.

Now stir into the coloured solution in the tall jars a tablespoonful of vinegar (add a little more if the solution does not fizz a little bit).

Now look closely at the mothballs, what do you notice?

What makes the moth balls move?

Explanation   ***
The bubbles formed when vinegar (an acid) reacts with the chemical (bicarbonate of soda) is carbon dioxide gas.

Bubbles of carbon dioxide stick to the moth ball and make it float to the surface. When the bubbles break at the top of the liquid the moth ball sinks again because it is heavy.

Acid is the Opposite of Alkali **

Sometimes when people get indigestion it is because they have over-eaten or over-drunk, a short while before. The stomach has to work very hard to try to digest all this extra material and, in doing so, sometimes produces too much stomach acid. This can be counteracted by taking stomach powders or tablets.

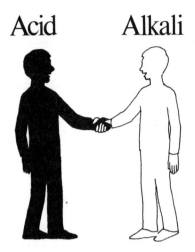

Acid   Alkali

You will need: Small quantity of either vinegar or a lemon
Small amount of bicarbonate of soda
Spoon and a glass

Sip a *small* amount of vinegar or suck a piece of lemon. What does it taste like?

Now *lick* a small quantity of bicarbonate of soda from a spoon.

What happens to the acid taste?
What happens on your tongue?
What are the fizzy bubbles?

Extension ***

When the vinegar (acid) reacts with bicarbonate of soda (alkali), they neutralise each other. When you get the right quantities of each they just balance and the tastes of each balance each other.

Try sipping the vinegar or lemon and then sucking a stomach tablet (like McLeans, Alka Seltzer, etc).

CHEMICALS AROUND THE HOME 65

Does the vinegar taste disappear? Was there any fizzing on your tongue?

The body always tries to keep the stomach in a balance and usually we do not need any extra stomach powders or tablets because the body has its own system of counteracting any excesses of acids or alkalis.

When an acid reacts with a carbonate, a gas, carbon dioxide, is given off and it is this gas that makes the fizzing sensation on your tongue.

Try adding the bicarbonate of soda powder directly with some vinegar in a cup and you will see this fizzing more clearly.

The Gas from Stomach Powders ***

You will need: Bicarbonate of soda
Vinegar
Birthday candle
Plasticine
Jam jar or tumbler

What to do
Place a blob of plasticine at the bottom of the jar and fix a candle in this. Surround the plasticine by a layer of bicarbonate of soda.
   Light the candle.
   Now carefully pour some vinegar down the inside of the glass until all the bicarbonate has been covered by the acid.
   Watch what happens to the lighted candle.

Extension ****
Try the same thing using Alka Seltzer instead of bicarbonate of soda and pour water carefully into the jar.
   The gas given off is carbon dioxide. Could you think of a use for this gas?

CHEMICALS AROUND THE HOME 67

Secret Messages   **(* *with help*)

Let's suppose that you are a secret agent and somehow you *must* send a secret message to another agent. How are you going to do it? You could use a special secret code or you could use invisible inks. We are going to use inks.

What you need: Piece of plain writing paper
1 lemon
Pen with a nib
Candle (or an electric iron for cloth)

What to do
Squeeze the juice of a lemon into a cup or glass.
   Use the juice as your invisible ink. Dip your pen into the juice and write your secret message on the paper (you might like to draw a map also).
   Let the writing dry. Can you see your writing?
   To make the writing appear either:
1   Warm the paper over the top of a candle, or
2   Iron the paper with a cloth iron, using only a *warm* iron.

Variations and extensions **
Invisible inks can also be made from a weak sugar solution, milk or a weak solution of bicarbonate of soda. All of these can be developed using heat from a candle or warming with an iron. Which is the best invisible ink?

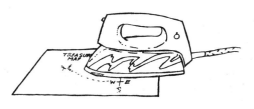

Changing Colours **

Introduction
Can you think of an animal that changes colour depending upon his background?

Do you know of any flowers that have different colours depending upon the type of soil that is present?

Some chemicals change colour depending upon the solution they are in. We are going to look at some of these.

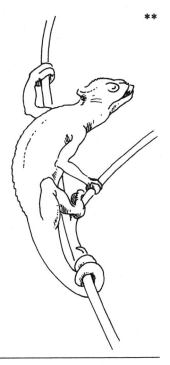

You will need: Some water after cooking potatoes or rice
A bottle of iodine and cotton wool pad (probably found in the medicine or first aid box)
Drawing paper
Narrow paint brush

What to do
Draw or paint a picture with cooled water after cooking potatoes (or rice). This will dry to give an almost colourless picture.

To colour the picture, gently wipe over it with a weak solution of iodine on a piece of cotton wool.

The first time you try this might not be the best, and so to get a good picture you might like to draw a few more pictures and colour them with the iodine solution.

CHEMICALS AROUND THE HOME 69

Extension ******

You could draw an animal which has black patches on it, say a panda, or a striped animal, say a zebra.

You draw in the black stripes in potato water then wipe over with iodine solution.

Trick question to ask your friends
Can you make the picture of a horse into a zebra without using a pen or pencil? Then you can use your iodine solution.

Coloured Flowers ***

You will need: Sheet of drawing paper
　　　　　　　Narrow paint brush
　　　　　　　Juice from beetroot
　　　　　　　Some ammonia
　　　　　　　Solution of bicarbonate of soda

What to do
Paint a picture of flowers with the beetroot juice. Let the picture dry. Wipe over the picture with either of the solutions of ammonia or bicarbonate of soda.

Extension ***
You could try to extract the colours from the petals of various flowers. If you want to see which flowers would give colours that will change colour in different solutions, then take a suitable flower petal, say from a rose, bluebell or wallflower and add a petal to vinegar (or acid) and another petal to bicarbonate of soda solution (an alkaline solution).

　　If the petals change colour then you know that these colours will change if you can only extract the colour.

　　Grind up the flowers with a few drops of water to extract the colour or gently warm the petals with a small quantity of water and use the solutions for your paintings. Dry the paintings and wipe over with (a) dilute vinegar solution, or (b) bicarbonate of soda solution.

Chemical Pictures   ***

First method

You will need: Some copper sulphate (you buy it as a solid but it can be easily dissolved in water)
Paint brush
Drawing paper

What to do
Paint your picture with a solution of copper sulphate. Allow the picture to dry.
   Now put your picture in a cardboard box with a small amount of ammonia solution in an open container.
   What do you note about your picture? Does the colour fade?

Second method

You will need: Cobalt chloride solution (in water). (You might have to buy this at the chemist's shop **and it is poisonous**)
Paint brush
Drawing paper

What to do
Paint your picture as in the other experiments.
   Dry the picture.
   Now, either warm up the picture on the paper near a fire, or high over a flame, or a lighted electric bulb, or iron the sheet with an electric iron, or else put the picture in a box containing ammonia solution in an open container.

ELECTRICITY

From the switching on of the light when we get up in the morning, to the switching off of the light as we snuggle down to go to sleep at night, electricity is part of our everyday lives.

Can you remember what happened when we had an electricity cut? The whole world seemed to have come to a stop. No lights, heating, cooking, traffic lights, radio or TV – they all went off!

But what is electricity? Can we see it? No. Smell it? No. Feel it? *Yes*, we can get an electric shock if we play around with mains electricity. This strength of electricity can kill. We are going to play safe and use low voltages. We shall use batteries as used in cycle lamps and those are safe.

The principles of electrical circuits are the same for low voltages and mains electricity. Let me emphasise again, **never, never play with mains electricity**.

Now let's get going!

ELECTRICITY 73

EXPERIMENTS IN THIS CHAPTER

Static Electricity *
Circuits 1*, 2**, 3**
Conductors and Insulators *
Question and Answer Board ***
Electromagnets *
Rob the Robot *(with help)
Making and Using Magnets **
Writing with Electricity **(with help)
Some Problems ****
Symbols used in Electricity

Star Rating Reminder
* Experiments that are relatively simple to do (if necessary, with help from parent or teacher).
** Experiments requiring a little thought and manipulative skills.
*** Experiments involving thought and reasoning powers and manipulative skills.
**** Experiments, problems and questions requiring a lot of thought and reasoning and often the designing of experiments, etc.

Materials You will Need:
Comb
Clear polythene sheet or cling film (1 foot long/30 cm)
Paper hanky
Coloured pens
A few Rice Krispies
Shallow cardboard box (like a paper hanky box) or polystyrene tray
Aluminium cooking foil
Sellotape
$4\frac{1}{2}$ volt flat battery
Paper clips
$4\frac{1}{2}$ volt bulbs (3 in all)
Sheet of thick cardboard or soft wood (about A4 paper size)
Plastic coated wire (only thin wire is needed, it can be bought cheaply in Woolworths) 2/3 metre long (2/3 yards)
Scissors
A 4″ or 6″ nail (10 cm or 15 cm)
A few paper clips
A thin piece of cork, or polystyrene or card (about the size of 10p coin)
Small plastic bowl about $\frac{1}{4}$ full of water
Sewing needle
Potassium iodide (very small quantity, can be bought from the chemist)
White hanky
Water after cooking potatoes or rice – when COLD
Drawing pins

ELECTRICITY 75

Some Electrical Terms Used

Terminals

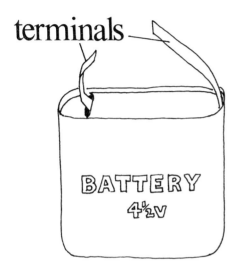

'v' means volts – A measure of the strength of the electricity; the greater the number the stronger the electricity.

Circuits
This is a complete loop of wire or foil, from one terminal through some items like bulbs, etc., to the other terminal.

If the 'circuit' is broken then no electricity flows. When the circuit is complete then electricity flows from one terminal of the battery to the other and the strength of the battery is used up.

76 SCIENCE FOR HANDICAPPED CHILDREN

Static Electricity *

I'm sure you have noticed that, after you have combed your hair, the hairs near the comb seem to be attracted to it. What causes this?

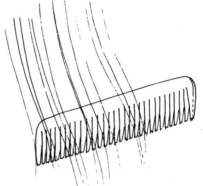

You will need: Some clear polythene or cling film
A paper hanky
Some coloured pens
A shallow cardboard box (paper hanky box)

What to do
Take the paper hanky and colour it with as many pens as you have got.

Rip a small piece of the coloured hanky about the size of half your finger nail and twist it to make a butterfly shape.

Draw a picture of flowers or a garden at the bottom, inside the box, with your coloured pens.

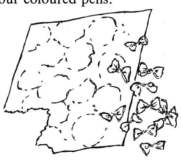

ELECTRICITY 77

Now put the 'butterflies' inside your box.

Cover the box with your sheet of clear plastic or cling film. Now gently rub the plastic cover with a dry hanky or any other material.

What do you see happening to your butterflies in your box garden?

The butterflies are attracted to the plastic top by electrostatic electricity. They move because the butterflies, with the same electrical charge on them, repel each other. They drop off the cover when the charge of attraction wears off.

Rub the top again to keep the butterflies flying.

Extension *

You could make a different picture backing to your box, say an underwater scene, and make the shapes like fishes and see them move around.

I have been told that a similar thing happens if Rice Krispies are used. You could colour the Krispies to look like beetles and ladybirds. Try and see.

Problem **

When the paper butterflies stick to the top, try rubbing the outside of the plastic top with another strip of plastic that has also been rubbed with a dry cloth. Does the strip repel the butterflies? Do the butterflies drop off?

Things with like or similar charge repel each other. Oppositely charged things attract. Negative (−) charge repels negative (−), but negative (−) charges attract positive (+) charges.

Circuits

It is important when dealing with electricity that it should obey your commands. We can turn electricity on and off by means of switches; these are like yes/no or on/off commands.

There are a number of electrical circuits around the house and a number of switches. Have you ever wondered what goes on inside the switch and where the wires go around the house? The experiments will *not* use mains electricity; we shall use low voltage and safe batteries and torch bulbs, but the circuits are the same.

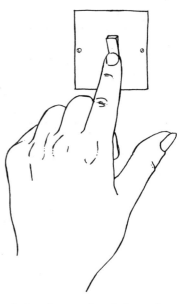

We are going to try a number of circuits on circuit boards – these are similar to those used around the house.

Circuit No. 1 * (with help)

You will need: Aluminium kitchen foil
Sellotape
$4\frac{1}{2}$ volt battery
Some paper clips
$4\frac{1}{2}$ volt bulb (the ones used in a cycle torch)
Sheet of thick corrugated card or soft plywood

What to do
Cut 2 strips of aluminium cooking foil – each about 1 foot (30 cm) long and about $\frac{1}{2}''$ (1 cm) wide.

Leaving $1\frac{1}{2}''$ (4 cm) free at the ends, sellotape the central 9" (23 cm) down onto the cardboard as shown in the diagram opposite.

First strip: Wrap one end around the screw part of the bulb and the other end around one terminal of the battery.
Second strip: Wrap one end around the other terminal of the battery and fold the other end upwards.

Put a piece of aluminium foil underneath the light bulb. This foil should be long enough for the 'upward folded' end of the second strip to touch – when the fold is flattened down.

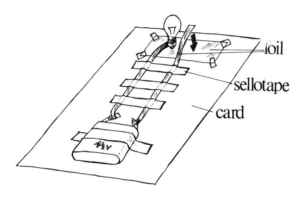

Now here's the big test:
If all your connections are well made, then when you fold the flap of aluminium foil downwards onto the strip under the light bulb, the light should go on.

Did it?

If it didn't then check to see if all the pieces of foil are connected tightly around the terminals of the battery and the bulb. Check to see if the bulb is touching sufficiently on the foil underneath it. You can stick a piece of sellotape over the bulb to keep it touching the foil beneath it.

The flap which you put down is like a switch and if you lift this flap the circuit is broken. What happens to the bulb?

What is a Circuit?
A circuit is when there is a complete loop of wire or conducting strip so that the electricity flows from one terminal to the other.

When the flap is lifted (or when the circuit is broken at any other point) then the flow of electricity stops and the bulb goes out.

Extension ***

Look at the side of the battery. Is there a + positive and − negative on it? Does it make any difference to your bulb if you turn the battery over and connect the terminals the other way round, ie the foil strip that was on the positive side now put on the negative side and vice versa?

In this experiment it doesn't make much difference which way around the terminals are connected but it does in some other experiments.

Circuit No. 2 **

Use the same circuit board as in Circuit No 1. We are going to add another bulb onto this circuit.

Put a second bulb in the circuit by wrapping the flap of aluminium foil around the screw part of the bulb and then place it firmly on the bare metal part of the aluminium sheet, similar to the first bulb. Make sure that the aluminium strip around the bulb does not touch the aluminium base.

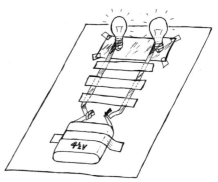

Some questions:
What happens to the light bulb?
Is it as bright as when using a single bulb?
Why do you think there is a change in brightness?
What has used up the brightness?
What do you think would happen to the brightness of the bulbs if we put a third bulb in the circuit?
Can you think of any way of connecting the two bulbs so that they are both as bright as the single bulb by itself? (See the next experiment).

ELECTRICITY 81

Extension **

When we have bulbs in a circuit, one joined onto another, we say it is in 'series'. When one bulb goes out then the other one goes out too.

Christmas tree light bulbs are sometimes connected in series. What happens if one of the bulbs breaks?

Circuit No 3 **

Connect up Circuit No 1, but now with aluminium strips put a third loop of foil from one battery terminal to the other side of the second bulb, and connect the strip attached to the bulb to the other terminal (as shown in diagram).

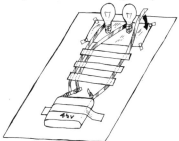

Questions

Are both the bulbs of the same brightness? Do they get the same amount of electricity from the battery?

We call these two circuits 'parallel' circuits.

Extensions ***

Can you make a circuit to connect three bulbs together all having the same brightness as a single bulb?

Make the circuit shown below. Is it in 'series' or 'parallel'?

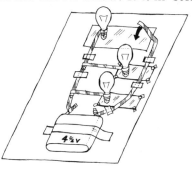

Conductors and Insulators

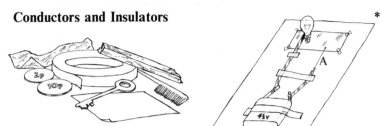

Use the same circuit as in No 1 (page 78).

Some materials, when placed in the gap marked A, in the diagram above, can cause the bulb to light up. These materials are called conductors.

Materials which do not conduct electricity are called insulators. Try some of the following:

Materials	Does the bulb go on?	Conductors or not conductors (insulators)
A 10p coin		
A copper coin		
A piece of wood		
A piece of plastic		
A piece of paper		
A piece of aluminium foil		
Any metallic object		
A piece of Sellotape		

Which materials would make good protectors against being affected by electric shocks? With which materials would it be unwise to touch anything electrical?

An electrician's screwdrivers or wire cutters always have handles covered with rubber or plastic. Can you say why?

Question and Answer Board   ***

You will need: A sheet of stiff card
　　　　　　　Aluminium strip or wires
　　　　　　　One 4½ volt battery
　　　　　　　One 4½ volt bulb
　　　　　　　Sellotape
　　　　　　　Paper fasteners

The circuit basically works like this:

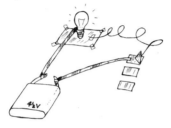

Follow the diagrams for both sides of the board, given on page 85.

When the strip of foil (or wire) from the question stud is connected to the correct answer then the light will go on. With incorrect answers no light will go on.

Here are some possible questions and answers for you to test the knowledge of your friends.

Question: What is the capital of the USA?
Possible answers:
　　New York
　　Boston
　　San Francisco
　　Los Angeles
　　Washington

You had better check the answer is correct before you connect up your circuit!

You can stick on different questions and answers to test yours and other people's knowledge.

Question and Answer Board

When you choose the correct answer this light will go on.

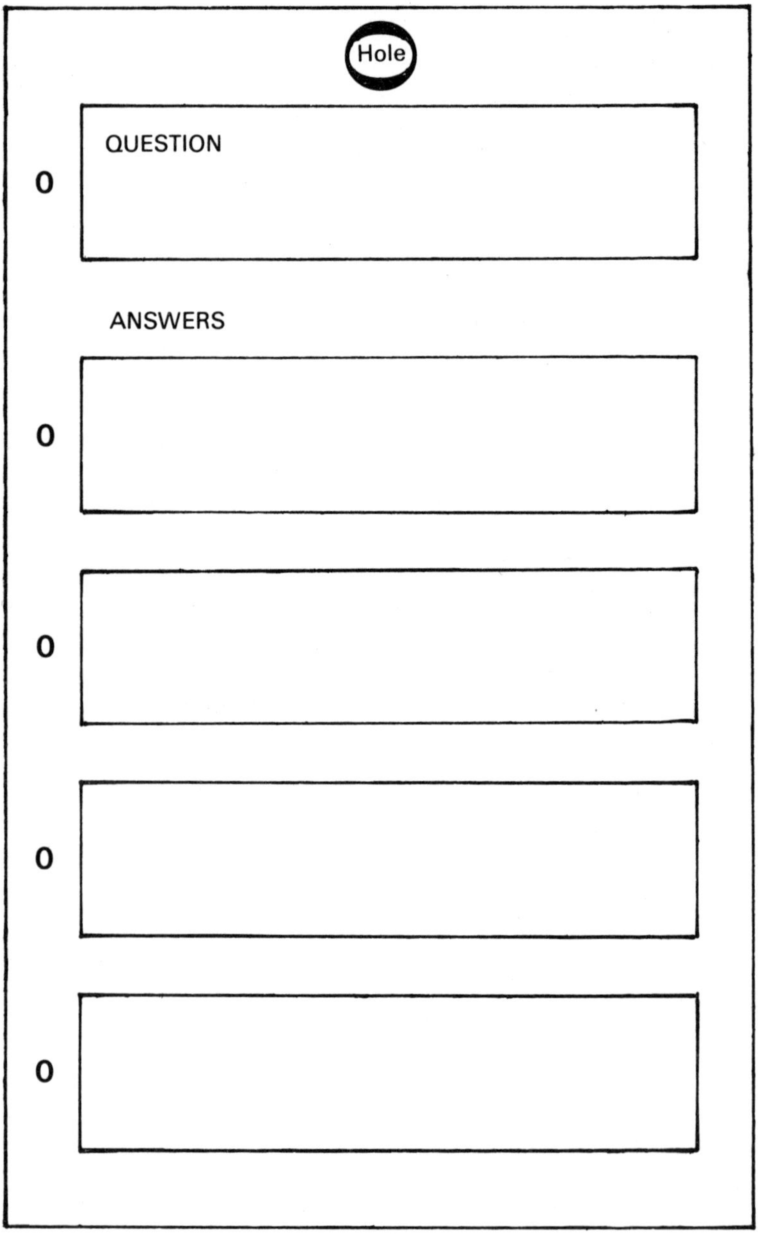

Hole

O QUESTION

ANSWERS

O

O

O

O

ELECTRICITY 85

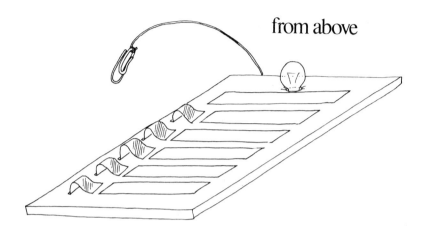

from above

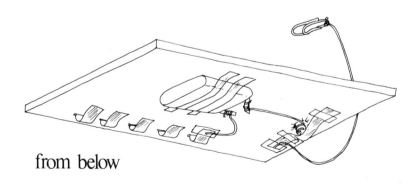

from below

86 SCIENCE FOR HANDICAPPED CHILDREN

Electromagnets *

We are going to make magnets using electricity just like the ones they use at a scrap yard.

You will need: 4½ volt battery
Length of insulated single core wire about 1 yard (1 metre) long (plastic coated wire)
1 nail

What to do
Wrap the insulated single wire around the nail in the same direction about 20 to 30 times.

Strip about 1 inch (2½ cm) of plastic from each end of the wire and wrap the bare wire once around each of the two battery flaps (as shown).

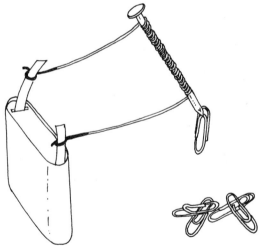

The electricity now spirals around the wire and around the nail. This causes the nail to be magnetised.

Now try to pick up some pins or paper clips with the end of the nail.

How many clips can you pick up?

When you disconnect the wire is the nail still magnetic?

ELECTRICITY 87

Rob the Robot * (with help)

This is a possible project for you to make a robot. I will give you a few suggestions but you might well be able to improve upon my design.

You will need: Cornflakes packet (for body)
Smaller box for head (say detergent box)
Tubes of cardboard or rolls of newspaper for arms and legs
Aluminium cooking foil (to cover robot)
Margarine tubs (for feet)
Odd pieces of cardboard (for hands)
Paper clips
2 bulbs
1 nail
Plastic coated wire
$4\frac{1}{2}$ volt battery

Explanations of the code letters are given overleaf.

a Circuits for eyes (see series circuits earlier)
b Holes for bulbs for eyes. Cover the holes with coloured sweet papers.
c Cover all the boxes with aluminium foil (it looks more like a robot then)
d Electromagnet (see the details for this in the previous experiment).
e To avoid wasting the battery all the time you can put a simple switch on his chest to switch on magnet and eyes (see earlier circuits).

Extensions   ****

1 You might buy a flashing bulb like the ones you put in Christmas tree lights. This makes the robot look very good.

2 Because Rob the Robot is covered entirely with conducting aluminium foil, perhaps you could use the surface to act as a conductor to the bulbs and put a battery connection to the coating of aluminium foil.

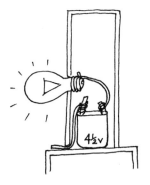

3 You could fix a small motor inside which makes something move, say the antennae on his head.

Making and Using Magnets

**

Many doors around the house make use of magnets to keep them closed – the refrigerator, deep freeze and some cupboards. We are going to use some of these magnets to make a compass.

You will need: Paper clip
Needle
Cork or piece of thick card or thin polystyrene

What to do
Use the paper clip to check where there are magnets around the house. The paper clip will be attracted to the magnet and hold on to it.

Were the magnets strong?

We are going to use the magnets to make other magnets.

Take the needle and smooth it over the surface of the part of the refrigerator door or door frame that has the strongest magnet in it.

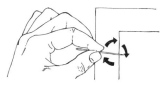

Smooth the needle in *one* direction only, about 60 times.

Now test whether the needle is able to pick up the paper clip or pin. If it does not attract a pin or paper clip, then repeat the rubbing another 40 times in the same direction.

Put the magnetised needle on a thin cork or piece of card and float this on the surface of a bowl or saucer of water.

Don't stir the water.

Does the needle float in one direction more than another?

Why?

What use could be made of this experiment?

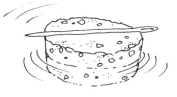

Is the North Pole the same as magnetic north?

You could also suspend the needle in a piece of paper on a long piece of cotton. Will this *always* point north? What happens if you approach your magnetised needle with an iron or steel object like scissors or a fork?

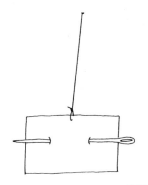

Extension ****

Put the loop of wire *near* (not touching) the needle. Does the electric current flowing through the wire have any effect upon the needle's position?

Do the same by putting the wire loop near the needle suspended on the cotton.

A further extension would be to put a few loops of wire all around the saucer and put the needle under the loops, then connect up the battery. Try different positions of the loop of wire and see if it has any effect upon the needle.

This instrument could be used to detect electricity passing through a wire.

It was in about 1819 that Hans Christian Oersted (pronounced URsted) noticed that a compass needle moved when he sent electricity through a wire near it. This discovery later led to the development of the electric motor, and think what our world would be like without electric motors... no electric machines, electric trains, etc.

ELECTRICITY 91

Writing with Electricity ** (with help)

You will need: Piece of cloth like a small handkerchief, or a piece of blotting paper
Some aluminium cooking foil
$4\frac{1}{2}$ volt battery and 2 wires
Some flour or starch (or water from cooking potatoes or rice)
Small quantity (not more than 1 gram or $\frac{1}{2}$ oz) of potassium iodide (bought from the chemist)

What to do
Take some water after cooking the potatoes (or rice) and when it is cool add about $\frac{1}{2}$ teaspoonful of potassium iodide crystals and stir in until it is all dissolved. Alternatively you can use water in which flour or starch have been soaked.

Take a white linen hanky or piece of blotting paper and soak it with the potato/iodide solution.

Put the hanky flat on the aluminium sheet and connect the sheet to the negative terminal of the battery.

Connect the positive end of the battery to the other piece of wire and use one end as a writing tool. Write your name on the wet hanky.

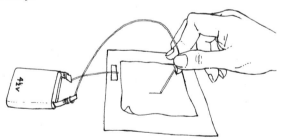

Explanation ***
The electricity liberates some iodine from the potassium iodide in the solution. This iodine forms a blue-black compound with the starch in the potato water. Don't leave the battery on too long as it wears it out.

Some Problems

The staircase problem ****

In your home you have probably got a switch at the bottom of the stairs which switches on a light, and when you go upstairs you switch off the light.

With your battery, strips of aluminium foil and a single light bulb, see if you can make a circuit which has two switches that work like the one in your home.

A possible answer is given below.

Further problems with electrical circuits ****

Can you make the following circuits?

1. A burglar alarm, so that when someone treads on a mat a bell rings or light is switched on (small $4\frac{1}{2}$ v circuit only need be used).
2. A switch that is turned on by a magnet.
3. When a toy car passes one point a light is switched on and when it passes a further point it is switched off.

Some suggested circuits are given on the opposite page but you might have solved the problem in a different way.

4. You might have noticed that when electricity passes through some of the wires in your experiments, the wires get hot. Why is this? Can use be made of this?

Answer to the staircase problem

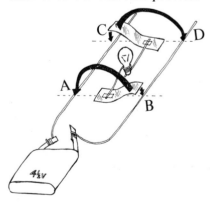

Try flaps in the following positions:

A	C
A	D
B	C
B	D

ELECTRICITY 93

Possible answers to the electrical circuit problems
1 Use separator strips of, say, thin foam rubber to keep the foil apart.

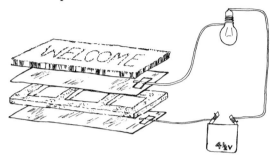

2 You can buy switches which are turned on and off by magnets. They are called Reed Switches.

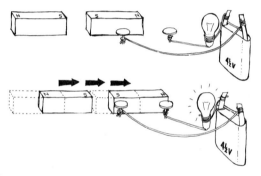

3

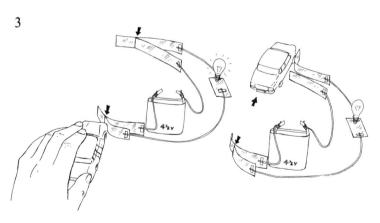

4 Electricity passes easier through some materials than others. When it is difficult, we say the wire has a resistance. In trying to get through the wire the electricity warms up the wire. Special wire can be made to convert a lot of electricity to heat. Such wire is used in electric fires. Other wires glow red or white hot when electricity passes through them; these are used in light bulbs.

Symbols Used in Electricity
Instead of writing out the exact shape of batteries, bulb, etc, in all circuits, a number of abbreviations or symbols are used. Here are a few of them:

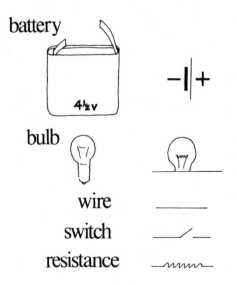

ICE AND WATER

Experiments with water are very common in various science schemes, but the use of ice is often neglected. The following experiments emphasise various scientific principles and hopefully introduce a sense of fun, interest and thought.

The experiments with ice, salt and water are designed to show that when a soluble impurity, like salt, is added to water, then this impure water freezes at a lower temperature; that is why we put salt on the roads in winter to keep the water from freezing, because salt water freezes at a lower temperature than pure water. When the sea freezes it is the pure water that freezes first and icebergs are usually fresh water.

The floating ice cube makes a good trick and causes a lot of discussion: will a full glass of water with ice cubes up to the brim, overflow when the ice melts? The principles here are to show that when water freezes it expands, but the volume of water the ice cube displaces when it floats is equal to the weight/volume of water of the ice cube. So . . . will the water level change?

Try and see.

96 SCIENCE FOR HANDICAPPED CHILDREN

EXPERIMENTS IN THIS CHAPTER

What is Ice? *
How Quickly does Ice Melt? **
Can You Watch Ice Melt? *
Lift a Lump of Ice with String or Cotton **
What Happens when you put Salt on Ice? ***
Can You Believe your Ice? ***
Arctic Water to the Sahara Desert **
Houses of Ice **
To Find What Effect other Objects Have on Ice **

Star Rating Reminder
* Experiments that are relatively simple to do (if necessary, with help from parent or teacher).
** Experiments requiring a little thought and manipulative skills.
*** Experiments involving thought and reasoning powers and manipulative skills.
**** Experiments, problems and questions requiring a lot of thought and reasoning and often the designing of experiments, etc.

Materials You will Need
Thermometer – which measures freezing point
Thermos flask
Pieces of cotton and string (approx 1–2 feet (30–60 cms) long)
Salt
Ice cube rack
Ice cubes
Plastic containers (margarine, ice cream containers)
Drinking glass or jam jar
Aluminium cooking foil
Plastic cling film
Cotton wool
Vegetable colouring
Plastic bags and ties
Liquid measure
Tray and bowls
Odds and ends (coins, etc.)

Hints and advice
It is probably best to do the experiments in, or near, a sink or on a large tray, or in a bowl. They are not messy but sometimes a bit wet!

Some experiments use blocks of ice and it is advisable to set these to freeze the day before you intend doing the experiments.

What is Ice? *

You need a refrigerator or freezer.

Make some ice by putting water into an ice-cube tray (or other container). Put your fridge at its lowest setting and place the container in the freezing compartment (or freezer).

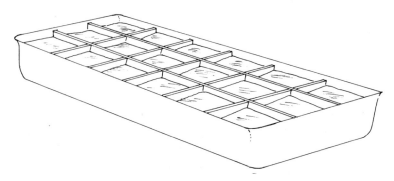

If you have a thermometer find out the following:

The temperature at which water freezes . . .°Fahrenheit
. . . °Centigrade (Celsius)
What is Ice?
Is it a Liquid, Solid, or Gas?
What is gaseous water called?

When your ice-cubes are frozen use some for the following experiment.

How Quickly does Ice Melt? **

Can we keep ice-cubes frozen longer by giving them 'overcoats'?

You will need: Ice-cubes
Some aluminium cooking foil, newspaper, plastic cling film, cotton wool, sawdust, thermos flask

What to do
Leaving *one* ice-cube unwrapped, wrap *one* ice-cube in each of your insulating materials or 'overcoats' (eg aluminium foil, cotton wool, cling film) and put one in a thermos flask.

Which ice-cube will melt quickest and which slowest? Can you guess? Write your guess below and ask other people for theirs:

Material	Names of people guessing			Actual result in the experiment
	Guessing time for melting			
	hrs. mins.	hrs. mins.	hrs. mins.	hrs. mins.
Foil	/	/	/	/
Newspaper	/	/	/	/
Cling film	/	/	/	/
Cotton wool	/	/	/	/
Sawdust	/	/	/	/
Flask	/	/	/	/

Which is the best insulating material? ....
Which is the worst insulating material? ....
Who was the closest with their guess? ....

Extension ******

Can you put an ice-cube in a suitable 'overcoat' in a *hot* oven so that it lasts – say – 2 minutes?

Can you put ice cream in the hot oven without it melting? Try making 'Baked Alaska'.

Baked Alaska ***** (with help)**

Ingredients: Block of ice cream
5 egg whites
6–10 oz caster sugar
Thin sponge cake
A few glacé cherries

Method: Turn oven to 475°F or Gas 8/9.

Arrange a block of ice cream on a thin sponge cake on a baking tin. Whisk egg whites until *very stiff* and pile over ice cream – completely covering the ice cream. Decorate with cherries.

Brown for 3–5 minutes in the very hot oven.
Did the ice cream melt? Why?

Iced Lollies *****

Add water to squash (half and half), pour into ice-cube tray and stick cocktail sticks in them. Freeze these. Your prize for doing these experiments: eat/suck your lolly. How can you keep your lolly in the room from melting?

Can You Watch Ice Melt?   *

You will need: Ice-cubes containing a few drops of vegetable food colouring (red, green, etc.) – fairly strong
3 glass jars (jam jars)

What to do
Freeze some coloured water into ice-cubes before you start the experiment. Fill the glass jars with water of different temperatures (ie cold, warm, hot).

Carefully drop a coloured ice-cube in each and watch what happens to the colours.

Which water temperature shows ice melting best?

Extension   **
Freeze ice-cubes containing other impurities such as sand, gravel, coffee, etc., and repeat above experiment and see if the impurities alter the speed of melting.

Lift a Lump of Ice with String or Cotton   **

You will need: Ice-cubes
  Piece of cotton and string
  Salt

What to do

Try to tie some cotton or string round an ice-cube. If you cannot do it in 30 seconds – try the following!

Put a few ice-cubes on a plate or tray. Put the end of cotton on top of a cube and sprinkle with a little salt. Leave a few seconds and try to lift the cube with the cotton. (Leave longer if it doesn't work first time)

When it does hold you can tie the cotton round the cube.

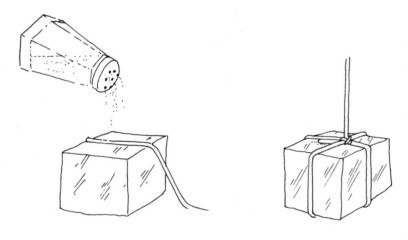

Ask your friends and family if they can 'lift' an ice-cube using only a piece of cotton or string without tying any knots in the cotton. Give them 30 seconds.

ICE AND WATER 103

What Happens when You Put Salt on Ice?   ***

You will need: A thermometer which measures freezing point
(you may have one in the freezer)
Ice-cubes in a bowl
Salt
Water

What to do
First, put the thermometer into the bowl with the ice-cubes and a little water (to dip the thermometer in), leave a few moments, then measure the temperature.

Add a few spoonfuls of salt and read the temperature after a few moments.

What happens? Can you explain the result?
Does water with added salt freeze at a lower temperature than pure water?

Extension   ***
Something to think about:
In the winter when the roads are icy the road workers sprinkle salt on the ice to melt the ice and snow.

Does the temperature of the ice/snow/water/salt road surface go up or down when the salt is added?

It is because the impure salt/water/ice mixture freezes at a lower temperature than ice alone. So salty water needs a much lower temperature before it freezes.

Can You Believe Your Ice? ***

There is a lot of ice at the North and South Poles. When pieces break off they are called icebergs and they float many miles until they melt.

If some icebergs lasted long enough to reach Britain and melted on our seashore, would people living there have to move to higher ground? Yes/No? Let's try an experiment to help you make up your mind.

You will need: Some ice-cubes
 Glass or jam jar
 Some water in a jug and a dry tray or bowl

What to do
Put a few ice-cubes in the glass and carefully fill to brim with cold water. (Do not spill).

Are the ice-cubes floating?
Is there more ice above or below
 the water?
Can you guess how much?
What will happen to the level of the
 water when the ice-cubes melt?
Will it overflow?
Will it remain the same?
Will the level go down?   Answer...................

Does water when frozen take up more, less or the same amount of room as before?

Repeat with a larger block of ice if you are not sure – try an ice-cream container or margarine tub, or even a plastic bag.

Would people living on a seashore have to move to higher ground if icebergs melted on our shores?

*Extension – to show that water when frozen
takes up more room* ***

You will need: Straight or narrow-necked yoghurt pots
Tops for them of grease-proof paper or foil
Water

What to do
Fill your container(s) to the brim with water and put on top(s) of thin card or paper. Put a dry container (eg a tray) to catch any overflow. Then freeze in a freezer.

Is there more ice now than there was water?

Have you seen frozen milk in bottles in winter?

Arctic Water to the Sahara Desert **

People in desert areas would love to have extra water. There is lots of frozen water at the Poles. Could very large icebergs be towed quickly from the Arctic to the Middle East?

To see how long lumps of ice take to melt

You will need: Water
 Containers – either plastic bags and ties or similar-shaped containers such as yoghurt pots, or large ice cream containers
 Several dry bowls

What to do
Measure quantities of water into the containers as follows:

1st container: ¼ pint or 1 measure
2nd   ,,   ½ ,, or 2 ,,
3rd   ,,   1 ,, or 4 ,,
4th   ,,   2 ,, or 8 ,,

Put them in the freezing compartment and leave to freeze

until all the blocks are frozen. (This might take all night, so continue this experiment tomorrow.)

Then, remove 'icebergs' from containers and put each into a separate bowl to melt – ensure the largest bowl is able to take over 2 pints.

Guess how long you think each 'iceberg' will take to melt. Ask your friends and family to guess also.

Iceberg with	Names of people guessing				Actual result of the experiment
	Guesses of the melting times				
¼ pint	hrs	hrs	hrs	hrs	hrs
½ pint	hrs	hrs	hrs	hrs	hrs
1 pint	hrs	hrs	hrs	hrs	hrs
2 pints	hrs	hrs	hrs	hrs	hrs

Can you say why you get this result? Did you guess correctly? Could you make a graph of the results. From your graph guess how long 3 pints of ice would take to melt.

24 hour 'iceberg'
Can you make an iceberg that takes all day to melt?
Do you think it would be possible to tow a real iceberg to the deserts?

Houses of Ice **

Eskimos used to live in igloos which were skilfully made from blocks of ice.

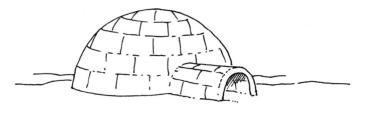

You will need: At least one tray of ice-cubes
Salt
A plate or tray to work on

What to do
Firstly – try to build a tower of ice-cubes. The cubes need to be evenly shaped.
How high did you get?

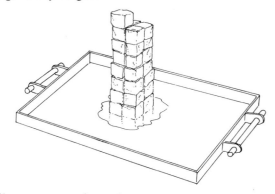

Secondly – put one ice-cube on your tray, then sprinkle a little salt on top of each cube as you add it to make them stick together.
How high did you get this time?
Try making a 'leaning tower' by putting salt on the tray beforehand!
Try to make an igloo – it is very difficult.

ICE AND WATER 109

To Find What Effect Other Objects Have on Ice **

You will need: Ice-cubes and various household objects
　　　　　　　Coins
　　　　　　　Spoons
　　　　　　　Salt
　　　　　　　Warm tap water

What to do
a Put ice-cubes on plate or tray and place various small objects on top of each (coin, aspirin, sweet, cork, rubber, sugar lump, etc.).
Does the ice melt more where the object touches it?

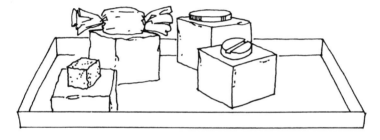

b Heat a coin or paper clip in very hot water (use a fork or tweezers to get it out) and put it on an ice-cube.
What happens?

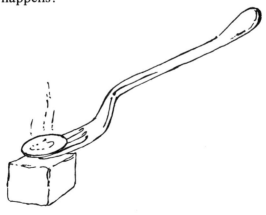

c Put an ice-cube in a tea-strainer and place an object on top. Gently pour hot water over.
   Does the object protect ice against the hot water?

d Pour a little salt onto an ice-cube.
   What happens?

e Put some crushed ice-cubes in a jar, add a little water, then add a few spoonfuls of salt. Leave a few minutes. What is forming on the outside of the glass jar? Why? Where did this come from?

FOAMS

These experiments with foams help to show that many foams contain air which can be squashed out and then reintroduced. Some foams are water repellent, others are not; some are good insulators and are used accordingly around the home. The insulation of a material often depends upon (a) the non-heat and sound conducting nature of the material and (b) the fact that air pockets are good heat and sound insulation (think of the air layer in double glazing units).

The floating and sinking experiments reveal the various properties of different materials when making boats. Boat experiments and the diver experiments show that something will float if it is over-all lighter than water (ie its density is less than that of water). The total weight of the materials plus air for the given volume must be lighter than the same volume of water if it is going to float.

EXPERIMENTS IN THIS CHAPTER

Foams around the House *
Does the Bathroom Sponge contain Air? *
Do All Foams Soak up Water? **
Squashability of a Sponge ** (with help)
Boats *
Divers *** (** with help)
Submarines ***
Swinging Water *

Star Rating Reminder
* Experiments that are relatively simple to do (if necessary, with help from parent or teacher).
** Experiments requiring a little thought and manipulative skills.
*** Experiments involving thought and reasoning powers and manipulative skills.
**** Experiments, problems and questions requiring a lot of thought and reasoning and often the designing of experiments, etc.

FOAMS

Materials You will Need
Pieces of foams found around the home, foam rubber of various types
Bath sponge
Measuring jug
Polystyrene packing or ceiling tile
Empty Bic pen case
Plastic squash bottle
Plasticine
Paper tissue
Andrews Liver Salts (or baking powder)
Plastic tubing
Small tins with lid (like syrup or cocoa tins)
Comb

Advice before You Start
It is probably best to do the experiments in, or near, a sink or on a large tray, or in a bowl. They are not messy but sometimes a bit wet.

Foams around the House

It is possible that the seat or chair you are now sitting on contains a soft layer of a plastic foam (we sometimes call this foam rubber).

Here are a number of different uses for various types of plastic foam. Look around your room or the house and see how many of these you have got and see if there are any I've missed.

Types of Use of Foams	Yes/No
Sponge in the bathroom or kitchen	
Foam cushions	
Foam pillows	
Cavity wall fillings	
Bathroom foam mat	
Polystyrene ceiling tiles	
Under carpet underlay	
Foam put in your shoes or trainers	
Foam linings to coats or anoraks	
Draught excluders on doors or windows	
Oven gloves	
Here's a chance for you to look for any I've missed:	

Does the Bathroom Sponge Contain Air? *

You will need: Bathroom sponge or car washing sponge
(or any piece of foam rubber)
Bowl of water (or use a sink)
Wide necked measuring jug or milk bottle

Now try this

1 Take a large bowl of water (or you could wait until you have a bath and do the experiment in the bath) and half fill the bowl with water.
2 Take the measuring jug and fill it up with water but do this under the surface of the water. Now turn the jug upside down so that it is full of water.
3 Take your dry sponge and put it under the water just underneath the jug. Don't let go the sponge, it will float inside the jug. Now gently squeeze the sponge with your fingers. What are the bubbles made of that float into the upturned jug? How much has the water in the jug gone down, caused by the bubbles?

4 Now remove the sponge and squeeze out as much water as possible with your hands and repeat the experiment. Does a wet sponge squeeze out as much air as a dry sponge? Explain this.

*Extensions to the experiments and
things to think about* **

A How much water will a sponge hold?
You could weigh a dry sponge on the kitchen scales and weigh it again when wet.

***

$$\frac{\text{Weight of wet sponge}}{\text{Weight of dry sponge}} = \frac{?}{?} = ?$$

B What happens when you dip a sponge completely in water and then hold it by the corner? What do you notice about the water in the sponge?

C Wet half the sponge, then take it out of the water and put the wet part at the top. What happens to the water?

D Hold a dry or squeezed sponge up to the light. Can you see through it? Now wet it and again hold it up to the light. Does this make any difference?

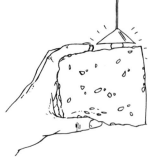

Do All Foams Soak up Water? **

Do all foams soak up water?
Look at the list you drew up on page 114 and think why some foams need to soak up water while other foams must be water repellent.

You will need: A few samples of foam plastics or rubbers
Bowl of water

Here are some hints on samples of foams:
1. A piece of ceiling tile.
2. A piece of foam cavity wall filling. (If you or your neighbours have had the cavity walls filled then ask for a piece of the foam that has usually escaped through small cracks into the loft.)
3. Piece of foam rubber sponge, etc.
4. A spare piece of underlay from a carpet.

Now try this
1. Use the bowl of water to find out which samples float and which sink.
2. Try to see which samples soak up water and which are only wet on the outside.
3. Look carefully at the surface of each of the foams and use a magnifying glass if you have got one.

To think about ***
Which of the foams would be good for making heat resistant cups for hot tea?
Which material is used for making wet suits for underwater divers and why?
Is it important that wall insulation materials must soak up or repel water?

118 SCIENCE FOR HANDICAPPED CHILDREN

Squashability of a Sponge   ** (* with help)

Although foams are generally made of soft material they can withstand quite high pressures and weights. The stair carpet or carpet in the room has an underlay or foam backing. All carpets have to put up with a lot of feet pressing on them.

The foam balls used in the game of tennis are hit with great force. What happens to their shape when hit?

Experiment: to test the squashability of some foams

You will need: A ceiling tile or piece of polystyrene sponge
Some spare carpet underlay
A foam ball

What to do
Take each of your foam samples and test it for its squashability.
1 When you press your finger into it, does the foam recover its shape?
2 Try squeezing the foam into as small a piece as possible. How long does it take to recover its original shape?
3 Can you flatten the foam? Try putting it in the jamb of the door and closing the door. The amount you can easily close

the door gives you a guide to its squashability. (Make sure your samples are all the same thickness if you want to compare them.)
4 Put a milk bottle full of water on a 'mat' made from each of your samples. Does the bottle leave a mark? Does the mark disappear as the foam gets back its shape when you take the bottle off. How long did it take?
5 Put some warm water in a bottle and use each of the foams in turn to see if you can feel the heat through the foams. Which ones are the best heat insulators?

Now answer the following questions:
1 Which materials would make the best insole in a shoe?
2 Which material would make a good oven glove?
3 Would a tennis game work as well with balls of polystyrene as with foam plastic?
4 Which material would make a good table mat to protect a polished table from being damaged with a hot teapot?
5 Would polystyrene be of any use as an underlay for carpets?

Extension: sound insulators   ***
Some foams are used as sound insulators to prevent the noise from one room being heard in another. Can you think of a way of testing which foams are good sound insulators?

Perhaps you could put a sheet of each of the foams chosen in front of the loudspeaker from a record player, radio or TV and see which cuts out most sound. You could also see how much you have got to turn the volume of the sound down so that you can just hear the sound with each of the foams in turn in front of the loudspeaker.

What uses can you think of for sound insulating foams?

Boats *

Some boats are made of wood and others of metal. We are going to do some experiments with both wooden and metal ones.

What was the Kon Tiki raft made of? Ask someone or look it up in a book.

You will need: Bowl half full of water (or you can do this experiment when you are in the bath)
Drawing paper
Aluminium cooking foil
Pieces of wood, like balsa wood
Polystyrene sheet or ceiling tile

What to do
Cut pieces of paper, wood, aluminium foil and polystyrene, all the same size, say 6" (15 cm) long by 3" (7.5 cm) wide.

Make a boat or raft shape with each material. You will have to fold the paper and aluminium foil up around the edges.

Carefully float these in your calm sea, your bowl of water. Check for leaks.

Which floats the best? Will any of these sink if water fills them, say in rough waters? Try and see.

Which boat can carry the most cargo?
Make a guess and see if you are right afterwards.

Take each boat in turn and very carefully put your cargo in the boat, say pennies or paper clips, one at a time until it just sinks. Make sure you load the pennies evenly along your boat. Count how many pennies make your boat sink.

	Number of pennies added before it sank
Wooden boat	
Paper boat	
Aluminium boat	
Polystyrene boat	

Which was the best boat in calm waters?

Would this be the best boat in rough waters? Try by doing the same experiment but this time make a few waves every time you add a penny; try not to get water into the boats.

Extensions **
Try a different shaped boat.
  Try a different material, stiff plastic, leather, canvas, etc.

Here's a quiz for you ***
  What is the hull of the *Queen Elizabeth* made of?
  Why do minesweepers in recent years have hulls of plastic?
  Why is a canoe made of glass fibre and not metal?
  The 'Owl and the pussy cat went to sea' in what coloured boat?
  What is the Plimsoll line on the side of a ship?
  Why were the fast moving frigates used in the Falkland Islands war made of aluminium and not iron or lead?

Divers *** (* with help)

Divers need air or oxygen to breathe under water. They must either carry the air with them or have it pumped down from the ship above.

When a diver goes down deep, then the pressure of the water above him compresses his body and the air in his body and he must be fed air under great pressure to keep his body and lungs from collapsing.

We are going to make a kind of diver.

Experiment to make a bottle diver

You will need: Large clear plastic bottle and top
              Clear plastic ball point pen case and top
                (no refill in it)
              Large bowl of water

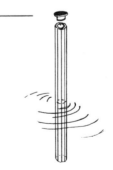

What to do
To make the diver:
Take the clear plastic case of a ball point pen, take the top off it (as shown) and hold the case upright in the bowl of water so that about ½ inch is showing above the water. Put the cap on

the case tightly so that you have captured an air bubble.

Carefully lift out the plastic case containing the air bubble and some water, and put it in the plastic bottle almost full of water.

The 'diver' (plastic case) should float in the water.

Put the top back tightly on the bottle.

Now

Squeeze the sides of the plastic bottle and watch the 'diver'.

What happens?

What happens when you stop squeezing the bottle?

Can you see why?

Why it works

Squeezing the bottle increases the pressure inside the bottle and this compresses the air bubble inside the 'diver' so that the air can no longer support the weight of the pen holder diver. So the pen sinks.

If the 'diver' doesn't work at first then it can be adjusted by putting a little bit more air in the 'diver' if it sinks to the bottom of the bottle when it is first put in *or* put less air bubble in the top if it doesn't easily sink when the side of the bottle is squeezed.

Submarines ***

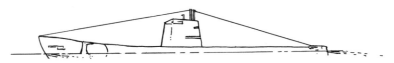

Submarines are ships that travel under the sea. They manage to stay under the sea by controlling the amount of air they hold inside them.

You will need: A small tin with a tight-fitting lid
A few inches of plastic tubing
A lump of plasticine
Paper tissue
Andrews Liver Salts

What to do
Remove the lid from your tin and then ask your parent or teacher to make two holes in it, just big enough for the plastic tubing to fit in them.

Cut two pieces of plastic tubing, one about 1½" long, the other just over twice the height of the tin.

Push the short piece of tubing into one hole and the longer piece into the other hole, then replace the lid on the tin.

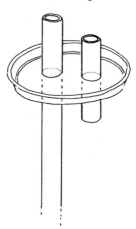

Your submarine will now look like this:

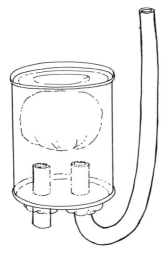

You must stick lumps of plasticine to the lid at the bottom of the tin. This seals the holes around the tubing and also acts as ballast (or extra weight). This will keep the tin upright. Be careful not to add too much.

To make your model work
Remove the tin from the water, remove the lid and put $\frac{1}{2}$ a teaspoon of Andrews Liver Salts into the centre of a paper tissue. Fold the tissue into a neat bundle. Slip this package into the tin. Secure the lid and then hold the submarine under water until it sinks.

Eventually the water will soak through the tissue. The liver salts will fizz and the gas will force the water out of the tin. When enough water is pushed out the submarine will float to the surface.

How can you make it sink again?

Swinging Water   *

You will need: A running tap
Some plastic rods (biro cases are OK) or comb

What to do
Use the tap in the kitchen or bathroom and turn the cold tap on so that you get a thin (narrow) continuous stream of water. You do not need to turn the tap on too high.

Now take a plastic rod or comb and rub the rod for about half a minute with a dry cloth (or your jumper), then place the rod close (but not touching) the stream of water.

Try the rod at different heights along the water stream.

Rub the rod after every try.

What happens to the water stream? Why?

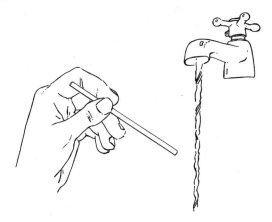

Did you try the rod on the other side of the stream? Did the same thing happen?

Try as many different plastic rods as possible.

Did all the plastic rods do the same thing to the water stream?

Try a wooden pencil. Did this deflect the water stream?

I wonder why plastic rods do this? Is there any connection between this experiment and the experiment that makes the butterflies jump?

CANDLES, LIGHT AND HOT AIR

This topic area covers experiments with candles, with torch lights and some with mirrors.

The experiments with a LIGHTED candle could, in some cases, be considered to be a hazard. They are very interesting and relatively safe and they can be used to give pupils confidence in working with potentially dangerous situations. It is, after all, a real life experience.

Even if the candle experiments are eliminated, many other interesting experiments are left to do in this section.

EXPERIMENTS IN THIS CHAPTER

Fire and Extinguishers ***
Candles and Flames – a few experiments **
Candles Burning in Jars ** and ***
Lights and Lamps *** (** with help)
Bending Light around Corners *** (** with help)
Writing with Mirrors *
Hot Air *
Temperatures Measured with Hot Air ***
Getting Dried Out **
A Project ****

Star Rating Reminder
*   Experiments that are relatively simple to do (if necessary, with help from parent or teacher).
**   Experiments requiring a little thought and manipulative skills.
*** Experiments involving thought and reasoning powers and manipulative skills.
**** Experiments, problems and questions requiring a lot of thought and reasoning and often the designing of experiments, etc.

Materials You will Need
Vinegar
Jug
Wide necked bowl, (eg food mixing bowl)
Some birthday candles
Plasticine or Blu-tak
Matches
Aluminium cooking foil
Short length of glass tubing
Bag of ice-cubes
Filler funnel
Few different sized glass jars (eg 1 lb, 2 lb jam jars)
Piece of cork or stiff card (to support some candles)
Cardboard box (shoe box)
Small electric torch
2 small mirrors
Stiff card or paper
A pair of glasses
Scissors
Writing paper
Length of cotton
Length (about 1 metre/1 yard) of clear plastic tubing
A milk bottle
Paper clips or Sellotape
Salt
Saucers
Egg cups
Methylated spirits
Cotton wool

Fire and Extinguishers ***

Fire is one of man's most useful tools, but when out of control it destroys everything in its path. This experiment may begin to explain to you how fire can be controlled by using the properties of some chemicals.

How are lighted candles affected by the reaction of bicarbonate of soda and vinegar?

You will need: A jug of vinegar
Bicarbonate of soda
A wide mixing bowl
Birthday candles of different heights
Plasticine or Blu-tak

Now try this
1 Secure the candles at the bottom of the mixing bowl with the plasticine or Blu-tak.
2 Put the bicarbonate of soda at the bottom of the mixing bowl, around the candles.

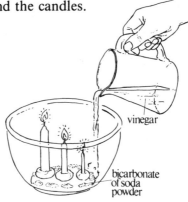

CANDLES, LIGHT AND HOT AIR

3   Light the candles.
4   Pour the vinegar gently down the side of the bowl on to the powder so it is just covered.
5   Notice which candle is extinguished first.
6   When all the candles have been extinguished try to relight them whilst still in the bowl.

Something to think about
Why is it difficult to relight the candles?
   What are the bubbles called coming from the bicarbonate of soda?
   How can you tell whether the bubbles of gas are heavier or lighter than air?
   An adaptation of this experiment is to set up the bowl and candles as above, but put the bicarbonate in the jug with the vinegar. Then pour the *gas* which is given off into the bowl, but do not let the liquid fall in. What happens to the candles? What could this gas be used for?

Explanation ****
The gas pours out of the jug and falls into the bottom of the bowl and creeps up the candles and eventually puts them out. Is this gas heavier or lighter than air?
   This heavy gas is called carbon dioxide.
   Sodium bicarbonate + Acid → Carbon dioxide
   (bicarbonate of soda)

$NaHCO_3$ + CH_3COOH → CH_3COONa + H_2O + CO_2
*sodium   acetic acid   sodium   water   carbon
bicarbonate   vinegar   acetate   dioxide
                                                        gas*

All these experiments put out the candle with carbon dioxide gas and many fire extinguishers either contain this gas or make it in a chemical reaction when the handle of the extinguisher is activated.

Candles and Flames ∗∗

Michael Faraday, an eminent scientist of the past century, gave a very famous lecture called 'The Chemical History of the Candle', and in it he explained a large number of basic ideas in science. We are going to look at some of the exciting and interesting experiments that can be done with something as simple as a candle.

You will need: A supply of candles (the ones used on birthday cakes would be all right, or the larger candles)
A box of matches
Some aluminium foil
Short piece of glass tubing

It is important in science to use all your senses to observe as much as possible. This experiment will test to see if you are a good detective.

What to do
Simply place a candle on a saucer or on a piece of aluminium foil. Light it and carefully look at the candle as it burns.

I want you to write down what you SEE.

There is a list given on the next page of some of the things you could have noticed. When you have watched the candle for five minutes and have

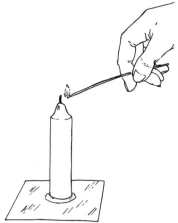

written down all your observations, then you can check to see what you have missed.

Did you notice all of these?
1. Height of flame; the shape of the flame.
2. Shape of the candle (it's a cylinder), height and width.
3. As the candle burns away it gets shorter.
4. Candle is made of soft white wax which melts.
5. Pool of candle wax at the top of the candle.
6. The pool of melted wax sometimes pours over the edge and goes solid again.
7. The top of the candle is not flat but has a dent in it.
8. The candle wick goes black and burns away slowly.
9. When the candle was lit with the lighted match you had to hold the match on the wick for a short while for it to catch fire.
10. The flame is made up of several colours:
    a bright yellow part
    a darker part
    a light blue part.
11. The flame flickers, particularly when near a draught or when gently blown.
12. Black smoke at the top of the flame.
13. The air above the candle is hot even a few inches above it.
14. The flame has quite smooth sides but is a bit ragged and wavy at the top.

15 The wick is white as it leaves the candle but black inside the flame.
16 The end of the burning wick has a red end to it.
17 The wick is not straight but curls over.
18 When the candle flame moves or the flame is blown, then the flame changes shape and sometimes the pool of hot wax falls down the edge of the candle.
19 The flame gives off a smell.
You might have noticed even more things.

It is important to have keen observations when doing experiments. If you didn't notice all these, look again at the burning candle and check off each observation to see if I was right.

Some of the observations you made will now be investigated further. This next experiment is:
To find out what is the fuel when a candle burns ****

What to do
Take a piece of aluminium foil about 3″ (8 cm) square. Cut a groove in the square and slide the foil across the lighted flame so that the groove slides over the wick.

Place the aluminium on top of the candle and leave it for a while and notice what happens to the flame.

Did the flame burn the same as before? Did it go out?
The aluminium foil stops the heat from the flame melting the top of the wax.
The solid wax does not burn but melts to liquid wax and the vapour coming from this is the real fuel.
When the candle goes out does the wick go out or does it smoulder?
Did it smell?

Extension   ****

You will need a piece of narrow *glass* tubing for this experiment which is designed to show that *near the wick* there is some unused wax fuel vapour.

This can be done using the apparatus shown in the diagram. The end of the glass tubing must be placed at an angle to collect the vapour going up the tube. Then light the vapour at the end of the glass tubing. That's Magic.

From these experiments, which of the following is correct?
The fuel of the candle is: a the solid wax
                                    b the liquid pool of melted wax
                                  c the wick
                                  d the vapour released from the melted wax.

What is the purpose of the wick?
Will the candle burn without the wick? Try burning a small piece of wax on its own and see.

Take a sheet of aluminium foil, put some ice-cubes in it and sprinkle some salt on the ice-cubes, wrap them up in the foil.

Hold the bag of ice cubes a few inches above the flame, with a clothes peg.

What did you notice was collecting on the surface of the ice bag?
Where does this come from?

Some explanations
The fuel of the candle is made up of hydrogen and carbon. When this burns (adds oxygen from the air) the hydrogen is burned to form steam which then condenses to water on the ice cold surface. The carbon burns to carbon dioxide gas. To find out whether this gas allows things to burn in it, collect the gas after burning, as shown below:

a   Lighted match.
b   A funnel held away from the flame so that it won't melt.

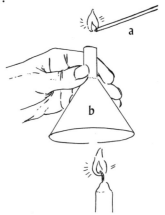

After the fuel has burned, the carbon dioxide is blown away. This gas can be directed onto a lighted splint or match. What happened to the match? Does this show that the gas supports burning?

Did you notice that on the aluminium ice bag there was also a black deposit? What would you call this? Soot? This black deposit is carbon, it is formed due to the fuel burning with insufficient oxygen around it; and instead of the fuel burning to carbon dioxide completely, some of it only produces carbon.

The hydrocarbon fuel of the candle, made up of carbon and hydrogen, burns in oxygen in the air

$$2H_2 + O_2 \rightarrow 2H_2O$$
hydrogen   oxygen   steam & water
$$C + O_2 \rightarrow CO_2$$
carbon   carbon dioxide

CANDLES, LIGHT AND HOT AIR

Candles Burning in Jars ** and ***

You will need: Some different sized jars
Deep bowl or bucket of water and some dishes or small bowls (like soup or dessert dishes)
A few candles
A few corks or small pieces of wood
Aluminium cooking foil

What to do
First experiment **
Collect the various sized jars and put them in order of size.

Now put your candles on a piece of aluminium cooking foil and set them alight.

Put a candle under each jar. Which candle will go out first? Try to explain why.

Second experiment ***
Now float a raft of the lighted candle on a piece of cork on the surface of water in one of the small bowls or a saucer.

Carefully put the jar over the top of the candle and place the jar on two pennies resting on the bottom of the bowl or saucer. Now watch carefully.
Try the same thing for each of the jars in turn.
What happens to the water levels?
Can you explain why?

Hint: Use a small amount of water in the saucer at first when putting the jar over the candle, later you can top the level up with more water.

Third experiment   ***

Ask a friend: Can a burning candle continue to burn under water?

Here's how to do the trick (You might have to practise this trick a few times to get it just right).

Put your candle floating on the raft on the surface of the water in bucket.

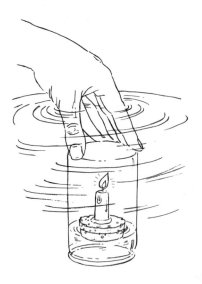

Take your biggest jar (used before) and place it over the candle, then firmly push the jar under the surface of the water. (You will probably have to practise this a few times.)

You will have to hold the jar at the bottom of the bucket with something, say a stick, or fill another jar with water and put it on top of the one containing the candle.

Lights and Lamps ***(** with help)

There are many sources of light which we see each day and you can easily name a few. The sun, stars, electric light, a candle, car lights, a glow worm!

Do you know how fast light travels? 186,000 miles each second!

You might have heard of a laser. This is a very intense and concentrated light. This very concentrated light was first made in 1960 and, probably, this was the first time ever that such a light had ever been seen in our Universe.

We are going to use torch lamps – not lasers.

Experiments with Light
You might have to do these experiments in a darkened room.

You will need: Torch that works
Cardboard box (like a shoe box)
2 mirrors (small handbag size – or two small mirror-tiles from DIY shop)
Stiff paper or card
A pair of glasses

What to do
1 Make some 'comb' type shapes of about 2½" × 2½" (6 cm × 6 cm)
   One with 1 cut.
   One with 3 cuts.
   One with 10 cuts.

Make a hole in the end of the cardboard box.
2 Put the torch in the cardboard box and turn it on. Put the cover on.
   It is best if you do these experiments in a fairly dark room.

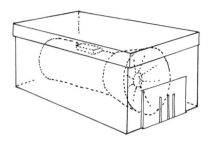

3   Place the three combs in front of the hole in the box in turn. What do the light rays look like?
4   Bounce the rays off a mirror.
5   What would happen if you used 2 mirrors arranged like this?

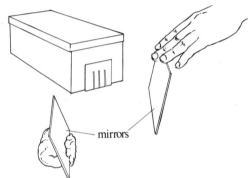

6   Put the mirrors at a series of different angles.
7   Try this, using comb number 3, with 10 holes:

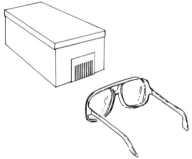

What happens after the light passes through the lens of the glasses?

Bending Light around Corners   *** (** with help)

You will need: Jam jar with screw top
              Torch that works
              Plastic bowl or sink

What to do
1. Fill a jam jar full of water.
2. Pierce the screw lid with a small nail.
3. Cover the jam jar with black paper.
4. Shine the torch through the end of the jam jar.
5. Pour the water out of the jam jar through the hole. Do this in a *dark* room.

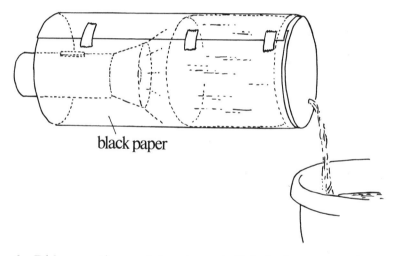
black paper

6. Did you notice anything about the light in the water and the bowl.
   Try to explain what you see.

The bending of light around corners can also be done using thin strands of glass rods. These are called glass fibres, and the technique is called fibre optics.
   You might have seen a lamp made of these fibres.
   Can you think of any other application of thin fibre optics?

142 SCIENCE FOR HANDICAPPED CHILDREN

Writing with Mirrors   *

You will need: Mirror
               Piece of paper
               Pencil or pen

What to do
Draw on the piece of paper a shape like a star. That's easy.

Now hold up your mirror (or ask someone else to hold it for you) so that by looking into the mirror you can see the star. That's easy.

Now use the pencil and trace around the shape you have drawn but you must *only look* at the mirror. That's not so easy.

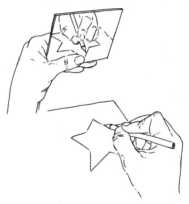

You can have fun with this. Can you do this experiment with your other hand?

Who does this shape the quickest in your house?

To think about   ***
Do you have to develop this skill?

Do you use these skills when combing your hair or shaving?

Don't be worried if you find this experiment difficult; so do I and lots of other people as well.

A dentist has to develop this skill as he sometimes uses a small mirror on the end of a rod to look at your back teeth, and sometimes to fill them also.

Hot Air *

Air is all around us and sometimes we feel it when it is windy, or when we blow out air from our mouths or whistle.

What causes the air to move and a wind or breeze to blow?

What causes a hot air balloon to rise and fall or the hang glider to be able to circle?

The reason is that air when it warms up gets thinner and tends to rise and the cold air falls to take its place.

We are going to look for places in the home where hot air rises.

You will need: A square of paper
Scissors
Pencil
Piece of cotton about $\frac{1}{2}$ metre long

What to do

Take the square of paper and draw on it a spiral pattern. It does not matter if it's not perfect. Cut along the line.

Thread the cotton through a hole in the centre and tie a knot in it to stop it pulling through.

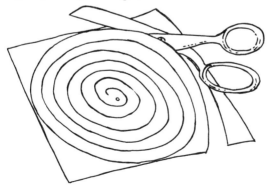

144 SCIENCE FOR HANDICAPPED CHILDREN

If you hold up your spiral it will form a shape like the one below:

Now, by holding the spiral on a long length of cotton above a place where hot air might be rising, it should begin to gently twist.

Try *above* a central heating radiator.
You might hang it above the reading lamp.
You could try a number of places.
You have discovered places where hot air is rising and so have found a mini wind source.

Extension ****

The spirals can be made in a variety of colours and used as Christmas decorations, and can be hung above the hot areas in the house, ie central heating radiators and lamps. This works well if you use aluminium foil also. Try other twisted shapes:

These upward flowing air currents are called thermals by the balloon and glider pilots and they occur when heat is rising from chimneys, factories or hot road or rock surfaces.

CANDLES, LIGHT AND HOT AIR 145

Temperatures Measured with Hot Air ***

You will need: Length of clean polythene tubing 1 yard
(1 metre) long
Milk bottle
Plasticine or Blu-tak
Paper clip (or sellotape)
Thick cardboard from cardboard box

What to do
1 Gather together the materials and pin (or sellotape) the tubing to the card as shown below.
2 Put a *small* quantity of coloured water in the tubing by carefully pouring it down the vertical open end.
3 Place the tube in the top of the bottle full of air and carefully seal up the bottle top with soft plasticine or Blu-tak.

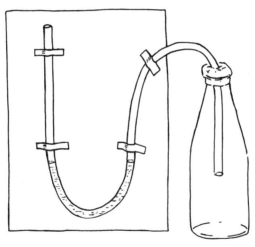

4 Note the position of the water in the tube. Now put both your hands around the bottle and watch the water in the U tube.
    Now remove your hands and allow the bottle to cool down to room temperature or even cool down with a damp cloth.
What happens to the water level?

Questions
What would happen to your U tube hot/cold meter if you put it:

   a   above the central heating radiator
   b   out in the cold weather or inside the refrigerator
   c   in a bucket of cold water or ice?

Try and see . . .

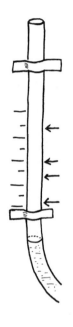

Extension   ***
Perhaps you could put a scale or rule alongside the tube and note the position of the water each day inside or outside the house. You could plot a graph of the values you measure.

This is your own thermometer.

Further extension experiment   ***
Instead of using a bottle of air, use instead a bottle of coloured water (water coloured with ink) and fill up the tube and bottle as far as the U in the tube.

Put this in the same positions as the air thermometer. Does the liquid expand as much as the air in the bottle?

What are the advantages and disadvantages with each type of thermometer?

Getting Dried Out

**

In our country things often get wet. How do they become dry again? Sometimes people deliberately make things wet in order to clean them. Why does use of a washing line, a spin drier or a tumble drier make the water come out of the clothes?

Find out what happens when a solution of salt and water is left in a variety of places.

You will need: Salt
Water
5 saucers
5 egg cups

Now try this
1. Mix tablespoons of salt with a $\frac{1}{4}$ pint of water, until no more salt will dissolve.
2. Put one tablespoon of the solution in each saucer and egg cup.
3. Leave one egg cup and saucer on a sunny window sill.
4. Leave the second egg cup and saucer in the bathroom.
5. Leave the third egg cup and saucer in the refrigerator.
6. Leave the fourth egg cup and saucer by a radiator which is switched on.

7   Leave the fifth egg cup and saucer outside in a windy place. Note what happens to the solution in each egg cup and saucer.

Something to think about ***

The air which surrounds us causes water to be removed from things. Air contains water vapour – that is, water which has evaporated to form a gas. If the air has a lot of water vapour in it, it cannot hold a lot more, but if the air has little water vapour in it, then water from ponds, plants, clothes and so on moves into the air. This leaves less water in the clothes, etc., so they become drier. Which do you think would dry first – a leaf in a steamy jungle or a leaf in a dry desert?

The temperature of the air is also important. Warm air can hold a lot more water vapour before it is saturated than cooler air. So in a warm temperature things generally lose water to the air more quickly than in a cool temperature.

Finally, if the same air stays around a cool wet object, the air soon becomes saturated with water vapour, but if the air is moving – if there is a wind – the air does not have a chance to become soaked before more air arrives to take its place.

Can you explain why the solution in the saucers evaporated more quickly than that in the egg cups? Why do spin and tumble driers make clothes dry?

Extension ***

Did you notice what the salt particles looked like when all the water had evaporated away? These are salt crystals; the slower the water evaporates away, the larger the crystals grow.

CANDLES, LIGHT AND HOT AIR 149

A Project   ****

Make a hot air balloon.

Take 4 large sheets of thin aluminium foil (you can sometimes buy coloured metal foil for packing presents) – metal foil is used as it is *non*-inflammable.

Join them as shown below left:

    Seal each seam with sellotape or sticky metal tape.

    Open out the one end to make as wide a circle as possible.

    Across the open end put *two* cross strands of aluminium foil.

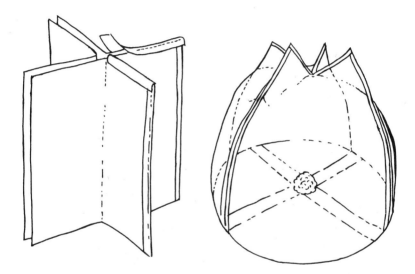

    Do the rest OUTSIDE.

    At the cross-over point of the strands of aluminium foil, attach a piece of cotton wool soaked in Methylated spirits (**not** Petrol) and carefully light. (Remember – do this outside.)

    The hot air will fill the bag above and, if you got the size and weight right, it should rise.

STRENGTHS OF MATERIALS

Many of the materials which we use every day are very strong, a piece of paper, a cotton thread, metal tube.

The strength of these materials is due to the way the fibres or molecules are held together in the materials. By a bit of ingenuity and clever designing even naturally weak materials can be made into strong ones. One piece of wood shaving is not very strong, but when compressed and mixed with a glue then we know that chipboard is very strong. The following experiments will try to reveal hidden strengths of some common materials. These include bridges, fibres, sticky tape and machines for using push/pull forces.

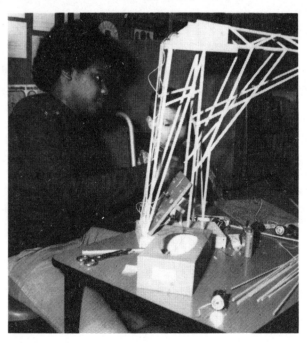

STRENGTHS OF MATERIALS   151

EXPERIMENTS IN THIS CHAPTER

Secret Strength *
Bridges on Windy Days *
Safe as Houses *
Straw Building **
The Geodesic Dome ***
Can a Piece of Paper Hold up a Brick? **
Make a Strong Building ***
Fibres, Strings and Ropes *
Woven Fibres *
Push, Pull Forces ** (with help)
Using Push, Pull Forces ****
Let's Make a Machine *
Balance Machine for Weighing Very Small Things ***

Star Rating Reminder
*   Experiments that are relatively simple to do (if necessary, with help from parent or teacher).
**   Experiments requiring a little thought and manipulative skills.
***   Experiments involving thought and reasoning powers and manipulative skills.
**** Experiments, problems and questions requiring a lot of thought and reasoning and often the designing of experiments, etc.

Materials You will Need
Plain writing paper
A number of 10p or other coins
Some match boxes or Lego bricks
Plasticine or Blu-tak
Pipecleaners
Drinking straws
PVA paper glue
Empty margarine tub and some sand
Scissors
Sellotape
Some potatoes
Stiff card
Rivet type paper clips
Pencils or felt tip pens
House brick or stone
Cotton
Yoghurt cups
Samples of various materials; nylon, denim, wool, etc.
Bucket

Note to Experimenter
Some of the experiments require rather fiddly hand manipulations and sometimes you need a third and fourth hand from a friend to help.

Secret Strength *

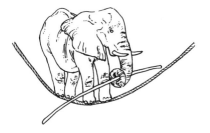

A single sheet of writing paper is not very strong and it is easily torn or ripped in half. But can you rip an *old* phone directory in half? Why?

You can snap a single piece of cotton but how many threads do you have to wind together before it is too difficult for you to snap easily?

Let's test the strength of a single piece of writing paper.

You will need: A few sheets of writing paper
                Some books to act as rests
                Some 10p coins

Now try this
Take two piles of books and between them place a piece of writing paper to act as a bridge.

See how many coins have to be added to cause the bridge to collapse.

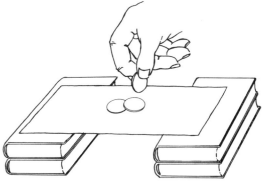

Use the same piece of writing paper and fold it like diagram 'A' below. Now test the strength using your coins.

Try folding another piece of writing paper in a different way, like diagram 'B' below, and then test its strength as a bridge.

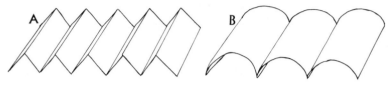

Shape of Bridge	Number of coins needed to collapse it
⟍⟋	
∧∧∧	
⊓⊔⊓	

Extension

Here are some other shaped bridges to try:

When you are out next time
1 Note especially the shapes of the bridges and sketch a few.
2 See how many are shaped like this:
and which have girders of various shapes in them:

3 If you ever go along a motor-way pay particular attention to the bridges, as they appear very often and there are some marvellous shapes.

You will be surprised if you count how many bridges there are on your journey.

STRENGTHS OF MATERIALS 155

Bridges on Windy Days *

Sometimes when it is very windy bridges which are over wide rivers are closed. Why is this?

There was one famous bridge in Tacoma in the USA which on one very windy day began to sway, and it eventually broke up completely.

Let's try an experiment to see the effect of wind on some different bridge shapes.

You will need: Some writing paper
 A lot of puff

What to do
1 Cut a strip of paper about 2″ (5 cm) wide and about 8″ (20 cm) long.
2 Make a bridge shape by bending the paper like this:

3 Place the bridge on a table and from about 2 feet (60 cm) away blow *underneath* the bridge. What happens to the shape of the bridge? Does the bridge blow away?
4 Try other shapes of bridge, eg:
 and see how they are affected by the wind.

What type and shape of bridge would you build over a motorway? What type of bridge over a river?

Safe as Houses? *

Look out of your window. Can you see a house? How are the bricks put together? Are they like this?

or this?

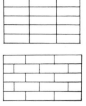

Perhaps some ways of building with bricks are stronger than others.

You will need: A set of Lego bricks or building bricks or matchboxes
Long piece of string 1 yard (1 metre) long
Blob of plasticine

What to do
You are going to be a builder and make two types of walls with Lego bricks or matchboxes.

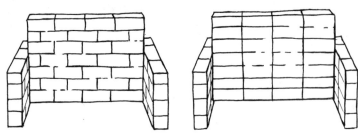

STRENGTHS OF MATERIALS   157

To test to see which wall is the best we are going to use a type of house demolishing machine.

It is a long piece of string with a heavy weight at the end, like a blob of plasticine.

Hold the top of the string as shown or tie it to the back of a chair and lift back the blob of plasticine, not too far at first.

Let the weight hit into the wall and see how far back you have got to lift the weight before the wall falls to pieces.

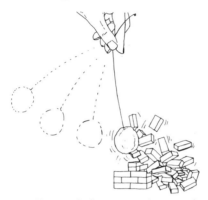

Did the two walls need the same destructive power or was one easier to break than the other?

I hope the builder of your house knew his job.

Extension   ***

You can use the demolishing machine much more accurately if you have a piece of card attached to the top of your string to see to what angle you lift back the blob.

Observations

Look at the tiling on the bathroom or kitchen walls. What pattern do they have?

Try drawing the patterns below and colouring alternate tiles. What does it look like?

Draw this for 10 rows and 10 tiles across. Can you see why your bathroom is tiled the way it is?

Straw Building **

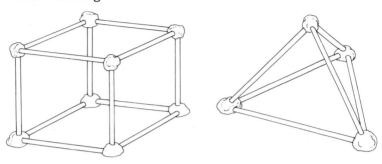

How do we make buildings strong?

You will need: At least 25 plastic straws (the same size)
Plasticine or pieces of bendy wire or pipe cleaners for the joints
Sheets of paper
Yoghurt carton
Coins or marbles
Piece of string

Experiment 1 To make a cube
Build a cube like the one above. Make your joints as strong and neat as possible using the plasticine or pipe cleaners.
  Test its strength by putting sheets of paper on top. What happens?

Experiment 2 To make a pyramid
Make the pyramid like the one above. Try hanging a yoghurt carton from the top. Do you think it is stronger than the cube? Fill the yoghurt carton with coins or marbles and try it.

Extension **
Can you add more straws to the cube to make it stronger? (Hint: look at the shapes of the pyramid sides.)

STRENGTHS OF MATERIALS 159

The Geodesic Dome   ***

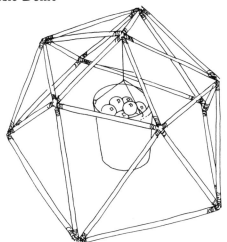

Now build your own geodesic dome. Make your joints as strong and as neat as possible using pipe cleaners or plasticine. The base is shaped like this:

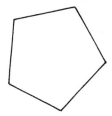

All the other shapes are triangles. Look at the picture above.
Test the strength of your dome by hanging a yoghurt carton from the top.
Fill the carton with coins or marbles to see if yours is a strong building.

Extension   ***
If you have room, make tubes out of rolled-up newspapers, bind them with sticky tape and make a big geodesic dome. Can you make one big enough for a den? You could cover the spaces between the tubes with newspaper triangles.
Can you find out which famous buildings are geodesic domes? There is one in the West Midlands Safari Park.

Can a Piece of Paper Hold up a Brick? **

You will need: PVA glue (white paper glue)
2 sheets of paper from an exercise book or big writing pad
Scissors
Empty margarine tub and sand or coins to test the strength of your platform
Sellotape

Make the supports
Fold one sheet of paper into 4 like this:

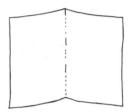  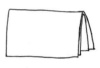

Open out and cut on the fold lines.
Now divide up each of these strips into 4 in the same way.
Wind these strips of paper around a pencil and stick the ends down with glue or sellotape to make sixteen strong tubes.

Take the other sheet and cut it in half. On one of these pieces, draw four lines across the paper an equal distance apart. Then draw four lines up and down the paper.
Put glue on to the end of your first tube and stick it standing upright where two of the lines cross. Stick the other tubes where the other lines cross.
Put glue on to the top of each tube and stick the last piece of paper on to the top of your platform.
Add coins to a margarine pot to test the strength of your platform.

STRENGTHS OF MATERIALS 161

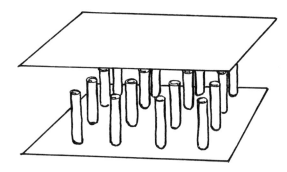

If it is very strong it might even support a house brick. Want to try?

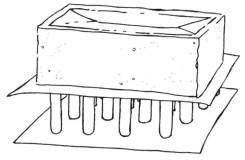

To think about
Look out for buildings supported by tubular columns.

Make a Strong Building ***

You will need: Some thin card
              PVA white paper glue
              Scissors
              Ruler
              Pencil
              Some weights

Cut 12 strips of card like this (measurements in cms):

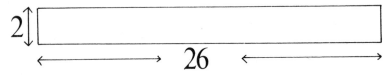

Now mark them off, using a rule:

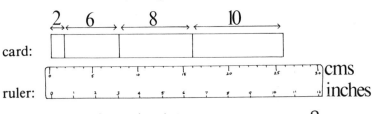

And fold them into triangles:

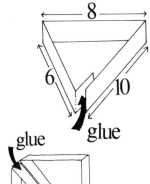

Stick 2 triangles together:

Watch which way round they are!
Do this for all the triangles.

STRENGTHS OF MATERIALS 163

Now stick these shapes here:

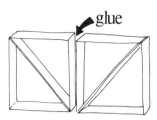

We next stick the pieces together like this:
(Make the bits of card or paper that go in between the layers by drawing round one of the pieces.)

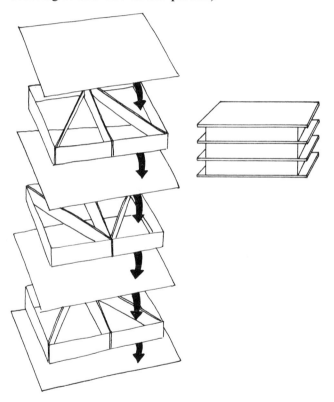

When the glue is dry, test your building's strength with weights, or pieces of stone. Are you surprised? Try turning it over? Will it support the weight of a housebrick?

Why is your building so strong? Is it strong in all directions?

Fibres, Strings & Ropes

We use ropes, strings and threads for many purposes around our home, from keeping buttons on to towing a car. What makes these fibres so strong?

Let's test the strength of a piece of cotton.

You will need: Reel of cotton or other threads
Small plastic bucket
Some potatoes

Now try this
Tie a length of cotton to the handle of a plastic bucket.

Tie the other end of the cotton to the back of a chair or door handle.

Carefully put potatoes into the bucket one at a time until the cotton breaks.

Now try doing the same using two lengths of cotton twisted together. Do you think you will need more or less potatoes?

Try three strands, twisted, and then try three strands plaited together.

STRENGTHS OF MATERIALS 165

Number of lengths of cotton	Single, twisted or plaited	Number of potatoes needed to break the thread
1		
2		
3		
4		

Extension ******

When you added the potatoes, you carefully added them to the bucket. Suppose you gave the thread a jerk, would it require more or less potatoes to break the thread? Why?

Try this
Fill the bucket as before and place one potato at a time into the bucket and let the bucket fall.

Does the thread break more or less easily?

Look at the various threads in the home using a magnifying glass, eg string, cotton, wool, etc. Are these single fibres? Can you easily separate the fibres? What makes string so strong?

Suppose you were a rock climber, what must you be sure of when buying a rope?

Suppose you were an AA or RAC breakdown mechanic and often had to tow other cars. What instructions would you make sure you gave the person being towed?

Woven Fibres

*

Sometimes fibres are woven together to make cloth and materials. These are much stronger than a single thread but they still wear out. Some materials fray while others fall into holes. Natural fibres, like wool and cotton, and also man made fibres, like nylon and terylene, all wear out eventually.

Experiment To see which materials last the longest

You will need: Some spare pieces of various materials, nylon, terylene, denim, wool
Block of wood or matchbox
Housebrick

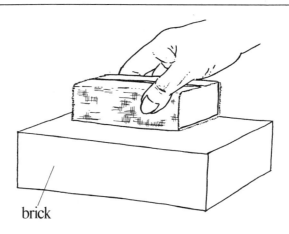

brick

What to do
Wrap your material around the block of wood or matchbox and either hold it tight at the top or stick a drawing pin into the block.

Count the number of times you need to rub the material on the brick to make a hole or signs of wear on the material.

Type of material	Number of times rubbed before a hole appears

Which would be the best material to make a set of working clothes?

Extension *

There are, of course, other ways of testing the materials. Can you think of any others? Devise, test and see which is the best.

You might see which is the best as a waterproof material by putting a few spots of water on the surface and watching to see how long it takes to soak in.

168 SCIENCE FOR HANDICAPPED CHILDREN

Push, Pull Forces ** (with help)

What you need: Piece of fairly stiff card about 8" (20 cm) square
Scissors
15 (or more) paper fasteners

What to do
Measure your piece of card and divide the width into 10 equal parts (if the strips might be too narrow then use two pieces of card). Make the widths about 1" (2½ cm) wide.

Make a pencil mark at the centre of each strip and also mark ½" (1 cm) from each end. Make a hole big enough for the paper clip to go through on all the points you have marked.

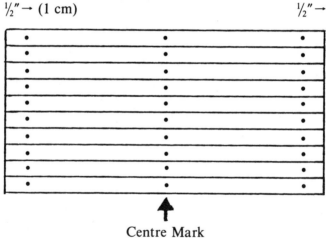

Centre Mark

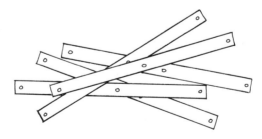

STRENGTHS OF MATERIALS 169

Cut out the strips. Arrange the strips in the shape shown below and put the paper clips through all the jointed points.

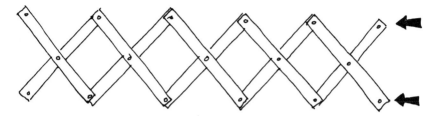

Push and pull the ends of your X shapes.
Now close up all the X shapes to make a shape.

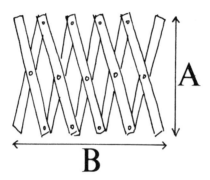

Measure the distances A & B.
Now stretch the X shapes to make A half as big and then measure the new size of B.
Can you think of any use for this X shape machine? Can you make use of this machine for anything?
Suppose you stuck a magnet on one end, what use could you make of it now?

170 SCIENCE FOR HANDICAPPED CHILDREN

Using Push, Pull Forces ****

You will need: The XX shapes from the previous experiment
Some plasticine or Blu-tak
Sharp pencils or felt tip pens

What to do
Take the cardboard strips and paper clips and make the following two shapes:

Experiment 1

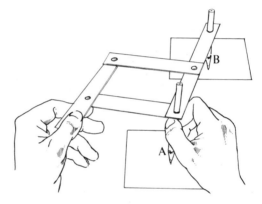

Experiment 2

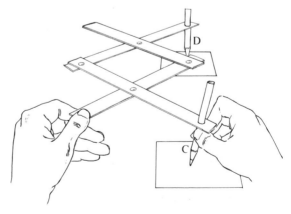

In Experiment 1 hold the shape with one hand as shown, and draw a square by holding pencil A.

Pencil B should be just touching the paper and should draw a shape on its own. What shape and size does B draw?

Do the same thing with Experiment 2. Draw a square with Pencil C. Does D draw the same shape and of the same size?

Try putting the pencils at other points and make other shapes with your cardboard strips.

What use can be made of these machines?

Extension **

Instead of joining the strips in the centre and at the ends try joining a few in the following way.

Make other holes here and use the strips for shapes as before.

A straight line drawing machine can be made like the one below.

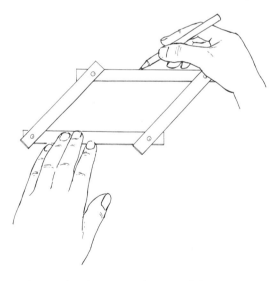

You can also make the same shapes with meccano or plastic meccano strips.

Let's Make a Machine *

What is a machine?
It's a piece of equipment that makes a job easier.

You will need: House brick or large stone
　　　　　　　Block of wood
　　　　　　　Measuring rule

Levers
Ask someone to get you the following: a house brick and a foot ruler plus a small block of wood.

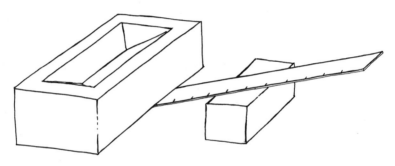

Place the ruler just under the edge of the brick and put the small block of wood about 6″ (15 cm) from the brick. Try pushing down at the end of the ruler. Try pushing the block of wood nearer the brick, and then later further away from the brick, to see if it makes it easier to lift.

Put two blocks under the ruler to make it higher. Does this make it easier to move the brick?

To think about ***

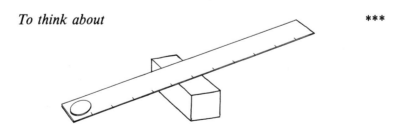

STRENGTHS OF MATERIALS 173

Balance a foot ruler at the centre on a narrow block of wood or support. Put one penny on the end of one side. How many pennies must you put on the other end to balance it?

How many pennies would you have to put half way along between the balance point and the end?

Try some other balance points and complete the box below.

Number of pennies on one side	Number of pennies on other side	Distance of pennies from centre of pivot

Can you see any pattern in these numbers?

A Simple Balance Machine for Weighing Very Small Things***

If you wanted to weigh something very small then a simple balance could be useful.

You will need: Thin piece of card or drinking straw
Stiff card or thin metal to make a U shape
Needle or knitting needle
Paperclips
Leaf

What to do

First make the balance as shown in the diagram overleaf.

Measure out a stiff piece of card to make a U shape (this is going to be the base of our balance). Pierce holes exactly opposite each other in the side of the U shape, as shown.

These will be supports for your needle or knitting needle point.

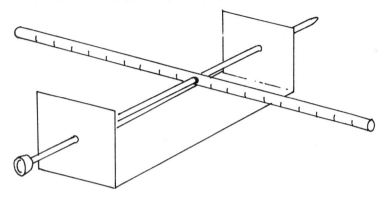

Measure out a thin piece of card or use a drinking straw. Mark lines on it every ½ cm (or every ¼"). In exactly the centre of the card (or straw) pierce a hole big enough for the needle to go through easily.

When put together the balance should look like the one above and the card or straw should be able to move easily up and down.

The balance arm should be exactly horizontal, if it is not, snip a small piece off the heavier side until it is horizontal.

Let's try our balance.

Put a paper clip, opened up slightly, on one side on the end mark, and put another clip on the opposite side at the same mark from the end. Do they balance?

Now let's see how many paper clips equal the weight of a small leaf.

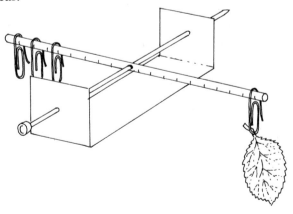

When it exactly balances, count the number of paper clips.

If you leave this leaf on your balance, will the weight change after a few hours? Try and see by leaving the apparatus set up and observe the angle of the balancing line.

Test the accuracy of your balance

Take two identical pieces of paper, say about one or two inches (5 cm) square. Place them on your balance to weigh them, one on each side; they should balance identically.

Now take *one* piece of paper and write your name and address on the paper. Now put it back on the scales and compare it with the plain piece. Is it the same? Does it change and show the weight of the ink in your name and address?

Extension ****

What could you use to find the exact weight of something in grams?

You could weigh on the kitchen scales 100 paper clips and then divide the weight by 100 to find the weight of one paper clip and so use your balance to weigh things in terms of paper clips and then multiply the weight found by the weight of a single paper clip, so giving the weight in grams.

SOME THINGS TO THINK ABOUT

The following collection of experiments is not all on one single theme but instead contains a set of interesting pieces of science. Many of these have a problem linked with them for you to think about.

SOME THINGS TO THINK ABOUT 177

EXPERIMENTS IN THIS CHAPTER

Coloured Windows * (with help)
TV Adverts ***
To Measure How Good Sticky Tape Is **
XL 20 ***
What is a Whammy-Diddle? ** (with help)
Happy Plants *
A Mathematical Twist ***
A Few Problems ****

Star Rating Reminder
* Experiments that are relatively simple to do (if necessary, with help from parent or teacher).
** Experiments requiring a little thought and manipulative skills.
*** Experiments involving thought and reasoning powers and manipulative skills.
**** Experiments, problems and questions requiring a lot of thought and reasoning and often the designing of experiments, etc.

Materials You will Need
Bandage
Tomato sauce or brown sauce
Containers of water, eg yoghurt cups
A selection of detergents
Bowl of water
Talcum powder
Rod of wood or pencil
Pin
Paper
Packets of cress seeds
Cotton wool
5 saucers
A plastic bag and twist tie
A straw
Scissors
Sellotape or stapler
Piece of writing paper
Empty tooth paste tube (wait until it is empty)
An aerosol
Cling film
Jam jar
Clear sheet of plastic or acetate sheet
2 pairs of polaroid sunglasses

Coloured Windows   * (with help)

In many old churches there are often very attractive stained glass windows. These give some very pretty colours within the church. This experiment shows how you can make some similar coloured effects using sellotape and polaroid sunglasses.

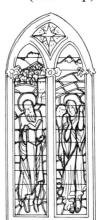

You will need: Roll of sellotape
　　　　　　　Frame of stiff card or blank frame from 35 mm slide frame
　　　　　　　Two pairs of polaroid sunglasses

Now you try this
You can overlap the strips of sellotape to make patterns.

Now put on a pair of polaroid sunglasses and hold your sellotape window up to the light. Can you see any coloured patches on the window?

Now hold the other pair of polaroid sunglasses close to your sellotape window as shown overleaf.

What colours can you see in your sellotape window? Here's magic: slowly rotate the sunglasses just behind your sellotape window. What do you see?

The explanation for this is rather complicated.

Extension   ***

You can make a number of different patterns and shapes of windows.

Do the number of layers of sellotape one on top of another make any difference to the colour?

Does one layer make the same colour as two layers, or three the same as two?

Further extension   ***

Take a single strip of sellotape or polythene bag and view it between your polaroid glasses.

Now stretch the sellotape. What do you see?

To think about

Have you ever noticed that on a sunny day when you have been in a car, and you have had polaroid sunglasses on, patterns appear on the windows and windscreen when the light is in certain directions? If you notice this again, twist your head around and see if the colours and patterns change.

TV Adverts ***

Adverts on TV – do they always tell the truth or give all the important details?

Sometimes adverts on TV say that scientists show that one thing is better than another, or scientific tests *prove* that this product is best. Let's do some scientific tests and see the truth of a possible advert.

You will need: Roll of bandage (or any other white material, piece of old hanky or sheet will do)
Tomato sauce or brown sauce
Containers of water, yoghurt cups and jam jars
Detergents, and if possible, one that contains a whitener and another which is 'biological'

What you do
Take about 2" (5 cm) of bandage and place it on some newspaper (it's going to be messy).

Now put some tomato sauce as evenly as possible along the whole length of the strip.

Cut the strip of bandage into, say, four pieces.

Piece number 1 we are going to keep to see what it looked like.
Piece 2 put in half a jar full of tap water.
Piece 3 put in some cold water into which you have already stirred up a detergent.
Piece 4 put in some other different detergent and water.

Soak Nos 2, 3 and 4 overnight or leave for 12 hours, then remove the bandage and wash under the tap. Dry all three in the same way and then compare which is the cleanest.

On TV sometimes they don't do the same thing to each strip. For example they might say soak one overnight and compare it with one that hasn't been soaked. That's not scientific. Then they make a comparison and say Brand Y washes best.

Try this *un*scientific comparison:

182 SCIENCE FOR HANDICAPPED CHILDREN

Make 4 strips as before.

Strip 1 keep for comparison.
Strip 2 put in hot pure water.
Strip 3 put in cold detergent water.
Strip 4 put in hot biological detergent and leave overnight.

Take them out, dry them and compare them. What's wrong with this experiment? Which of these washed the best?

To Measure How Good Sticky Tape Is ✱✱

You will need: Roll of sticky tape
Some different surfaces like piece of glass or mirror, piece of card, piece of plastic or floortile, etc.

What you do
On each material measure a length of 2″ (5 cm) and stick the tape firmly to it.

Measure another 2″ (5 cm) and then wrap the sticky tape around a 10p coin a few times and cut off the extra tape.

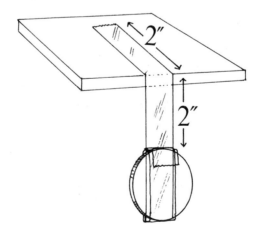

Guess which surface will hang on to the coin stuck to the tape for the longest time.

Now try each in turn using some lengths of tape, etc., and time how long it takes for the tape to come off.

If it's too long a time to measure easily, then use a shorter length stuck to the surface of the material and possibly more coins. Whatever you do, keep it the same for all the surfaces.

If you were the advertiser, which surface would *you* use on TV to show how good your sticky tape was?

XL 20 ***

'Compound XL 20, the new additive, works wonders. It makes dirt disappear.'

You will need: Bowl half full of water
　　　　　　　Washing up liquid
　　　　　　　Talcum powder (or pepper from a pepper pot)

What to do

Let's pretend that the washing up liquid contains the magic ingredient XL 20.

Half fill the bowl with clean water.

Lightly sprinkle the 'dirt' (talcum powder) over the surface of the water so that it *just* forms a thin covering.

Now carefully and with great pomp and show drop one spot of ordinary washing up liquid in the centre of the bowl of dirty water.

Does the dirt disappear? What happens to it?

Is it a 'scientific' fact that XL 20 removes dirt?

Keep your eye on the TV adverts so that they will not fool you by using unscientific experiments.

What is a Whammy-Diddle? ** (with help)

This is the name of a piece of simple apparatus that my father showed me when I was a child.

Let me give you the instructions on how to make one and then you can see if you understand how it works.

You will need: Rod of wood about as long and as narrow as a pencil (you can use a pencil if need be)
Sharp knife
Small piece of paper or thin card
Pin

What to do
Take the rod and cut a series of small grooves in it. See the diagram below.

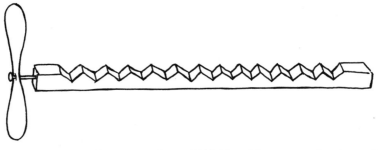

Cut a propeller shape about $1\frac{1}{2}''$ (4 cm) long out of a piece of thin card or paper. Stick a pin through the centre of the paper propeller and push the pin into the end of the grooved rod.

Here's the big test!!

Hold the free end of the whammy-diddle between two fingers and now rub the grooves with another piece of wood (pencil) or a biro in *one* direction only.

If you hold the rod carefully (try your fingers in different positions) and rub it with a number of quick rubs, the propeller vibrates and sometimes begins to turn.

If you change your finger positions, with care you can cause the propeller to turn in the opposite direction.

SOME THINGS TO THINK ABOUT 185

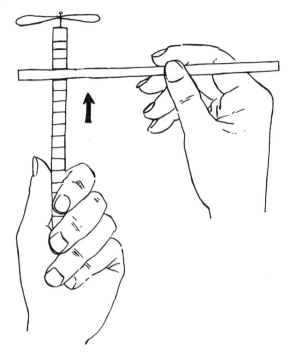

This is not an easy experiment to do and you have to have *patience* and probably try different holding positions with your fingers and different rates of rubbing with the rod.

When you find the positions to hold the pencil to make the propeller turn, remember it, then you can use the whammy-diddle as a magic stick to try with your friends.

You can say, 'Watch as I make the propeller turn by rubbing in the *forward* direction. What will I have to do to make the propeller turn in the opposite direction?' (Often they will say rub in the opposite direction, but actually you rub in the same direction as before but change the place you hold the grooved wood.)

Happy Plants

In the summer there are plants with many leaves and flowers which you can see on wasteland, roadside and even growing on walls. What makes them grow so well in summer but almost die in the winter?

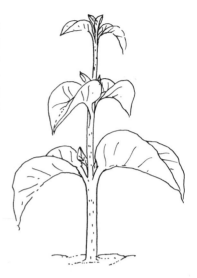

Let's find out the conditions in which seeds will grow best.

You will need: Packet of cress seeds
Cotton wool
5 saucers
Plastic bag and twist tie
Straw

Now try this
1. Spread a layer of cotton wool on each saucer.
2. Sprinkle some cress seeds on each piece of cotton wool.
3. Leave one saucer of seeds on a window sill.
4. Moisten the other saucers of seeds with cold tap water.
5. Put one saucer in a transparent polythene bag. Use a straw to suck out the air then seal the bag with a twist tie. Leave on a window sill.
6. Put the third saucer in the refrigerator.
7. Put the fourth saucer in a dark drawer where no light can get to it.

8 Leave the fifth saucer on a window sill with water on the cotton wool.

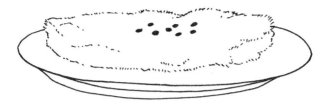

9 Leave all the saucers for one week, but dampen the cotton wool regularly on all except the first saucer.
Where have the seeds grown best of all?

Something to think about ***
What happens to the climate and daylight in winter which stops most plants from growing?

Look carefully at a piece of ground near your home. Write the names of all the plants which grow there. What happens to them in winter? Do the same kind of plants grow in the same place during the following summer? Why might this be? Why do you think lots of plants grow in the hot damp jungles near the equator, but hardly anything grows in the cold frozen arctic circle?

A Mathematical Twist

***

Look around your room and see how many pieces of furniture or other objects have twists in them. For example, can you see any twisted string, plants that twist around a stick, patterns on furniture?

Do the twists make the structures stronger or are they for decoration only?

You will need: Piece of paper, say one foot (30 cm) long
Scissors
Sellotape or stapler

What to do

Measure lines about 1" (2½ cm) apart along the longest length of your paper, as shown.

Cut along these lines to make some long strips.

SOME THINGS TO THINK ABOUT

Activity 1
Take one strip and join up the ends to make a circle. Does this circular strip have an inside and an outside? How many edges does it have?

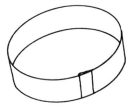

Activity 2
Take a second strip and twist it once then join the ends to make a loop.

Now run your finger around the outside surface of your loop and you will find something funny happens. Has this loop got an inside and an outside?

How many edges does it have? Run your finger along the edges and find out.

Here's one extra trick
Take your twisted loop and cut it right down the centre like this, so that you cut it in half. Cut along the dotted line.
What have you found out about the two loops you have cut?

Would the same thing happen if you cut the circle you made first of all?

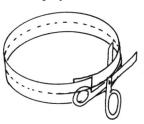

Extension
You could try twisting the strips a *few* times and joining them up into loops and then cutting these as before. Is there any pattern in the number of loops on cutting, and the number of twists?

A Few Problems   ****

1 When all the toothpaste has been used from a tube and the tube has been flattened, is there a way to re-inflate the tube?

2 Let's read what is written on the side of some household containers:
Aerosols: Why does it say 'Store in a cool place' and 'Take care not to puncture the can'? What makes the stuff come out of the can? eg shaving soap, hair spray, etc. Spray some of these materials on a piece of unwanted glass or old mirror and watch carefully. Does anything evaporate away?

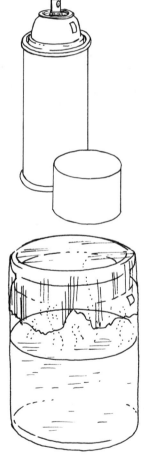

3 Half fill a jam jar, cup or glass with very warm water either out of the hot tap or from a kettle (take care not to scald yourself).

Enclose and seal the top of the jar, etc, with cling film and smooth down the edges to make sure it is sealed.

Let the water cool. Explain what you see. What will happen if you put this jar in a jug of hot water?

SOME THINGS TO THINK ABOUT 191

What happens if you put a half filled jar of cold water, sealed with cling film, into a jug of hot water?

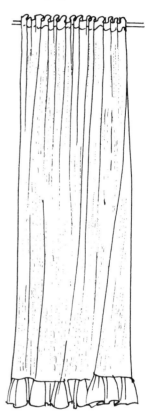

4 Have you got any net curtains in the house? Have you ever noticed the patterns formed when you look through a few folds of these curtains?

What causes these patterns and why do they change when the folds are slightly moved?

Only a very small movement in the curtains causes a change in pattern, is there any possible use for this?

Here are some other patterns for you to try.

Draw a circle about the size of a dinner plate with a thick felt tip pen and draw a circle at the centre about the size of a 10p coin.

Now draw thin lines from one side of the circle to the other – as shown overleaf.

Then draw lines between these lines, and keep doing that a few times more.

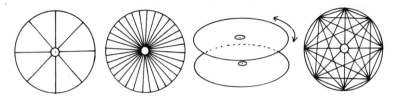

Draw the *same* pattern on a clear sheet of plastic using exactly the same size.

When you have completed this, place the plastic sheet exactly on top of the other sheet. Now slightly move the top plastic sheet and note the effect of this.

Extension
Try drawing other patterns on the two sheets to see if they interact also.

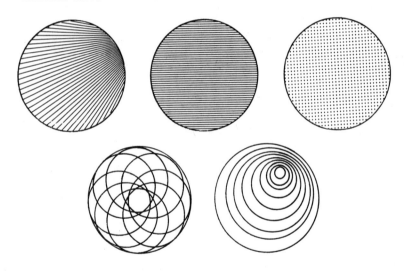

APPENDIX I

MAINLY FOR TEACHERS, INSPECTORS AND EDUCATORS

In Part 1 of this book it was stated that our main purpose in writing was to promote science for pupils who do not normally learn the subject or who have often been thought incapable of tackling the necessary physical and mental manipulations. Pupils who are able to cope with examination orientated science courses, in either the separate sciences or integrated science, have a number of science schemes and syllabuses available. The examination boards, whether CSE or 'O' level,

have syllabuses to suit many approaches to science, ie Chemistry or Chemistry with Physics, Integrated Science, Human Biology, Biology with Chemistry, Biology with Physics, General Science, Agricultural Science, Applied Science and Technology, Food and Nutrition; these are just a few of the syllabuses listed in one examination board's 'O' level syllabus booklet.

The chance for teachers to submit their own syllabus at 'O' level or, more popularly, to local CSE boards, opens up opportunities for a more individualised set of courses. It might be suitable for one school to construct a syllabus entitled 'Science for Life' which could include the school's own brand of philosophy towards CSE science. The flexibility of syllabuses is such that limited grade CSE's courses can be written which could cater for even the lower ability ranges and slow learners. The mode 3 assessment procedures can help the teacher to slant the course work and activities marks towards the children's abilities and not their disabilities.

The examination boards are most accommodating towards helping handicapped pupils overcome the barriers of the examination system and, providing the boards are consulted far enough in advance, a satisfactory arrangement for each candidate can be made. The applications usually need to be accompanied by appropriate medical certificates and specifications of the degree and type of handicap. It might be possible in some circumstances to have oral examinations in some science subjects. *Schools Council Examination Bulletin Number 21* provides details.[6]

There has been much written and talked about, in recent years, concerning the necessity for a 'core' curriculum of science. Often this has meant a central collection of established stalwarts of scientific concepts which no self-respecting scientist can do without. In some cases people have put forward scientific syllabuses and ideas which have tried both to lay a sound foundation of science and also to contain material which is relevant, innovative and interesting. It is probable that, for the slow learner, and for people who have missed a lot of schooling due to hospitalisation, *relevance* and *interest* are the most important pegs on which to hang the syllabus contents.

A great deal of excellent work has been financed by various agencies during the past decade or more, and very attractive sets of syllabus and work schemes have been made available to the teacher. The next few pages will give a brief summary of some of the possible schemes upon which a relevant science curriculum can be drawn. It might be that one single scheme, book or collection of worksheets will be suited to a school. What often happens is that the teacher modifies some experiments from one scheme and merges them with experiments from another, and thus synthesises his or her own science curriculum which is suitable for the individual needs of a particular school.

The schemes chosen have been analysed for their possible use with handicapped children and in some cases have also been tried and tested in schools.

Chronological Ages

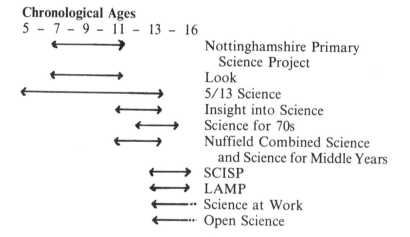

The selection of British projects[7] outlined above contains a wealth of material which can be scrutinised at leisure at your local teachers' centre or College/Poly/University library. Often science advisers employed by the local education offices will tell you which schools are using which schemes. A good indication of whether a scheme is appropriate is to talk to the people using it. Full details of the published schemes are given in the references on page 212.

The science projects mentioned above, and many others, have been developed with various categories of children in mind. The teachers and research workers who put these schemes together took into account both the current educational theories and child psychology. The exact philosophy and chosen approach can be more easily unravelled by reading the teachers' guides to all these schemes. One recent book entitled *Towards a Science of Science Teaching* by Shayer and Adey[4] would make interesting reading for anyone wanting to see how Piaget's work has been used in developing a number of currently used school science schemes.

The Look Scheme
These beautifully presented work sheets, together with two teachers' guides A & B, were designed for all ability ranges of children in the primary schools. Worksheets of Scheme A covered children of chronological age 7-9 and Scheme B for 9-11 year-olds. The teachers' guide shows how the worksheets can be used to supplement topic work or linked with other ways of working within the primary science classroom. An apparatus guide is given together with suggested ways of approaching science. No apparatus is supplied with this scheme but it follows a similar pattern to the Nottinghamshire Primary Science Project and many experiments could be 'boxed' for future use, or the apparatus stored centrally (say on a science trolley or in a set of drawers) and withdrawn as required.

The Look and Nottinghamshire Primary Science Project could both be used with children of higher chronological age but lower developmental age. Both schemes have been tried in handicapped schools and used as high up the chronological age as 13, but with children of lower mental age. They are very motivating schemes which encourage the children to think and enjoy *doing* science.

Nottinghamshire Primary Science Project
This project, based at Trent Polytechnic (1979-82), was developed to encourage a practically based and first hand science scheme for children between the ages of 7-11. It was developed by a team of advisers and lecturers (Cyril Gilbert,

John Carter, Alan Jones, Jim McLaren, Shirley Dean) who supplied ideas and experiments for 18 teachers (employed by MSC) to try in local schools. The ideas were repeated and evaluated a number of times in local schools by experienced practising teachers in the 40 trial schools. In order to encourage the teachers to start on the work with the minimum amount of disruption to the classroom routine, 180 science experiments were produced on worksheets, together with an equivalent number of totally self-contained experiment boxes.

All 180 experiments were scrutinised for use with handicapped pupils and were categorised using the PULTIBEC System. Not all 180 experiments were thought to be appropriate for all categories of handicapped pupils, but the vast majority were. Care was taken with language and concept development of the children when designing the work.

Insight into Science
This series of topic areas, extensively backed up with well prepared worksheets and teachers' notes, is geared to the early years of the secondary school, covering the chronological ages of 11–13. The material is presented in such a way that it can be used by older children with a lower developmental age, and no offence is given by the material being too 'young'.

The scheme has been well thought out and in any single topic area the work cards of activities are numbered. There is a code in the teachers' notes indicating which cards are suitable for everyone (core) and which ones would be suitable for the slow learner or brighter children. If a circus method is used, whereby the children in the class have to do, say, five activities out of eight or ten, then the most appropriate worksheet can be chosen for any particular child.

This scheme is used quite widely in Britain. In schools where the handicapped child is integrated into the comprehensive school, suitable cards can be chosen which are related to his abilities. A number of the topic areas have been modified by the Trent Polytechnic Science for Handicapped Pupils Project and some categorisation has been made. For example, the sheets containing bunsen burners for heating have been modified to show electrical heaters (which are safer than open flames but still produce the desired effect).

The scheme has a very useful set of pupils' readers which are well prepared, easily read and clearly illustrated. The teachers' book contains copyright-free masters for assessments, extra class work, resources for experiments and guides to other resources needed.

The topic areas (each contains about 80 double sided A4 cards) include:
- Sound
- Materials
- Colour and Light
- Electricity and Magnetism
- Forces
- Air and Heat
- Growth and Development
- Invertebrates
- Measurement
- Environment
- Sorting things out
- Water

Science for 70s (and Science for 80s)

This very effective set of integrated science topics contains work which was originally designed for children in Scotland between 12 and 14 years old, but the material has often found favour and use in the English schools also. The publications contain two teachers' guides, two pupils' manuals (fully and colourfully illustrated) and a collection of tear off work sheets. The units are not sequential and any topic can be used in isolation.

The teachers' guide shows how the scheme can be used in various contexts, for around the central core of material there is also a high and low level alternative route.

Unit 1 Introductory Science
 2 Looking at Living Things
 3 Energy – The Basic Idea
 4 Matter as Particles
 5 Solvents and Solutions
 6 Cells and Reproduction
 7 Electricity

8   Some Common Gases
9   Making Heat Flow
10 Hydrogen, Acids and Alkalis
11 Detecting the Environment
12 The Earth
13 Support and Movement
14 Transport Systems
15 Electricity and Magnetism

5/13 Science
This project, designed for the stimulation of science in the age range indicated by the title, produced a number of well prepared books for teachers, but no books or material for pupils. There is also a handbook covering the philosophy of the scheme, entitled *With Objectives in Mind*, and a booklet suggesting ideas for infant science called *Early Experiences*. The whole scheme is based upon the Piaget model of child development and the books for junior pupils at Piaget Stage II cover a wide range of topics including: Science of Toys, Wood, Metals, Ourselves, Minibeasts.

The Piaget Stage III books cover topic titles which lead the children into the realms of abstract thinking.

This particular science scheme is for the enthusiast, a person who is prepared to put together a syllabus based on the ideas. The latter can easily be woven into various topic areas and the writers have thoroughly indicated what specific objective the activities should exemplify.

The major central aim of this scheme is 'to develop an enquiring mind'.

LAMP Project
The Association for Science Education, an association of mainly secondary science teachers, sponsored a scheme which would realistically produce material concentrating upon the needs of the 'Least Academically Motivated Pupils', ie the slow and often the reluctant learner. A group of science teachers put together a series of topics which they knew would be a success with 14–16 year-old pupils. The topics are in typed form and are purposely made as cheap as possible. The material contains experimental sheets for pupils (which can be photocopied or

duplicated without fear of arrest), background information sheets for pupils and also helpful ideas and apparatus lists for teachers. There is an overall teachers' guide which explains the purposes of this scheme.

The topics covered are:

The Motor Car	Fabrics
Gardening	Heating & Lighting a Home
Electronics	Pollution
Photography	Materials
Fuels	Health & Hygiene
Flight	Space & Space Travel
Food Science	Paints & Dyes

There is little or no emphasis on difficult or mathematically related concepts.

This scheme has the great advantage of being cheap, realistically written, clearly illustrated (although without elaborate coloured graphics) and tested and written by practising teachers. The topics are written to provide a teacher resource and intended to be of particular use to inexperienced teachers or often non-science specialists who are attempting to teach the less academic 14–16 year-old using a topic approach. Many schools are using this scheme as a core syllabus for CSE mode 3, often with limited grade ranges. The topics can be used independently and are *not* examination orientated.

Schools Council Integrated Science Project – Patterns
For children able to continue with science concepts beyond the developmental age of 13, and for a school wanting to present an integrated science project based on a practical discovery approach to science, SCISP is most adaptable. A series of books is available: the teachers' handbook contains the philosophy and thinking behind the scheme; the pupils' books, entitled *Patterns 1, 2, 3 and 4*, backed up by corresponding teachers' books 1–4 and Technicians' Manuals 1–4, are a comprehensive set of publications. There are also a number of background readers which supplement individual topics. The syllabus is designed so that, while a complete set of concepts and experiments is given, the teacher is encouraged to replace any outdated or unsuitable sections with his or her own material.

If the course if followed for 'O' level or CSE in its entirety then about one fifth of the school timetable time is required for the 13-16 year-olds. The scheme is written and examined so as to give a *double* 'O' level or CSE pass. The scheme is flexible enough for modification for a mode 3 CSE syllabus or limited grade CSE, or can be used as a resource of good ideas.

The idea of integrated science does satisfy the need for a relevant and balanced coverage of the sciences.

Science at Work
Science at Work is a series of 18 units of science-orientated work and activities for 14-16 year-old pupils who are generally drawn from the less able subgroups in the school.

Each unit has a pupils' book and a teachers' guide pamphlet. The teachers' guide outlines the realistic aims of the course and gives detailed teaching notes on the topic under discussion, including apparatus lists, new words and the answers to questions covered in the pupils' books. It produces copyright-free record sheets for experimental data and suggests some specimen post-unit questions which could be used in tests, homework, examinations, etc.

The colourful pupils' books provide clear (but sometimes wordy) instructions for the practical investigations. The book of experiments becomes progressively more difficult towards the end and so the slowest and weakest pupils might not be able to cope with the later experiments. The work has been checked for suitable language level and reading ages.

The units can be used individually or inter-related and formulated into a mode 3 CSE syllabus.

A number of experiments and topic areas are very suitable for the handicapped pupil.

The topic areas cover:
  Fibres and Fabrics
  Photography
  Food and Microbes
  Forensic Science
  Pollution
  Cosmetics
  Dyes and Dyeing
  Gears and Gearing

Building Science
Domestic Electricity
Body Maintenance
Electronics

Open Science
This science scheme was researched with funds from the Schools Council and tested in some 30 secondary schools. The aim of the project was to produce interesting science for 14–16 year-olds who were normally in the lower bands of academic achievement and who were often poorly motivated, had a restricted concentration span and limitations of vocabulary and language skills (probably the bottom 30% of the ability range). Teachers' guide and pupils' work units and record sheets are produced. Topic areas covered include: Electricity in the home; Finding out about machines; Fire; Grow your own; Keeping going; Life spotting; Pollution; Safe eating; Science at home; Snaps and circuits; Starting and Stopping; Structures. Each topic area has separate well thought out experimental activities. The teachers' notes on each activity provide background information. This scheme provides tried and tested activities suited to most classroom needs.

SAVI/SELPH

S	Science	S	Science
A	Activities for	E	Enrichment activities for
V	Visually	L	Learners with
I	Impaired pupils	P	Physical
		H	Handicaps

This scheme of science activities was developed in the USA and was tested in classrooms for five years before becoming commercially available.

The scheme has been specifically designed for pupils in the upper ages of junior schools but is probably suitable for lower secondary also, particularly for retarded learners. The experimentally orientated scheme 'de-emphasises' disabilities but capabilities are emphasised. All sensory and manipulative abilities are brought to bear on the subject under investigation.

The material developed was not envisaged as a total

curriculum but as a set of enrichment activities, and these were designed for flexibility of use. 'The classroom teacher should feel comfortable to mould the instruction period to suit the needs of students and other classroom activities. Some teachers may find SAVI fits their needs once a month in an all-day exposure while other teachers may use SAVI daily for 10–15 minutes. SAVI lends itself to both these users.'

The materials available are 'modules' – a collection of self contained activities. Instructional material is provided for the teacher in pamphlet form. Some topic areas are:

Scientific reasoning: Structures of life; Communication; Mixtures and solutions; Environments; Measurement; Environmental energy; Kitchen interactions; Magnetism and electricity.

Special 'multisensory' apparatus is used, including adaptations for blind pupils.

Nuffield Combined Science

This science scheme was devised for 11–13 year-old pupils and can be adapted for the whole ability range.

The guiding principle of the scheme was to produce a source of ideas and materials to allow individual teachers, or schools, to devise their own courses. Teachers' guidebooks, a guide to apparatus and equipment, and pupils' activity books, are available.

There is no discrete syllabus for this scheme and teachers are encouraged to select and adapt the material included in the books.

Sections of the scheme include:

The World around us	Water
Looking for patterns	Small things
How living things begin	Earth
Air	Insects
Electricity	Energy

The exact sequence of these topics and individual activities within the topics are left to the teacher, although a suggested pathway is given.

The scheme is intended to provide a good foundation of sound science for all children between 11–13 and to be a starter for any subsequent science course.

In 1977 the Nuffield Combined Science scheme was extended to produce very well prepared study cards and practical work cards. These are arranged in themes, each of which has a booklet for teachers, which includes apparatus, visual aid, material list, teaching hints and sequences.

Themes include:
  Movement
  Out of doors
  Estimating and Measuring
  Colour in Nature

Possible Curriculum Choices

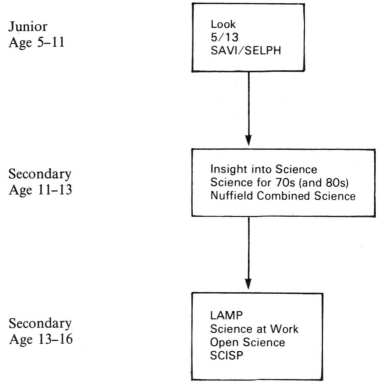

Junior
Age 5–11

  Look
  5/13
  SAVI/SELPH

Secondary
Age 11–13

  Insight into Science
  Science for 70s (and 80s)
  Nuffield Combined Science

Secondary
Age 13–16

  LAMP
  Science at Work
  Open Science
  SCISP

As time progresses other science schemes will appear on the market and can be scrutinised for their possible use with handicapped pupils.

APPENDIX II

BASIC CATEGORIES OF SOME HANDICAPS

Spina Bifida
The majority of children with spina bifida are paralysed from below the waist, have no feeling in their lower limbs, and are incontinent as a result of damage to the spinal cord.

Spina bifida coupled with hydrocephalus gives rise to many learning problems. Fine motor control is often poor. Children have difficulties when writing and manipulating objects. Hand and eye co-ordination are poor. Number concepts are slow to develop. There is a lack of body awareness and of position in space.

These children often have difficulty in keeping a place in a book and in transferring information from a text to an exercise book. Front, behind, left and right are easily confused. Depth perception is distorted, giving the children a floating feeling.

Muscular Dystrophy
This is a genetic condition which involves a weakness and degeneration of the muscles of the body and is a progressive disease leading to deformity and death. Eventually the child's limbs need to be placed by the teacher into required positions. Gross motor movements are usually impossible by the age of 13 years. Little pressure can be exerted by the hand due to lack of muscle tone. Objects need to be handed to the child who can write, but with difficulty. Electric wheelchairs are often relied on for mobility.

Cerebral Palsy
The degree of cerebral palsy may vary from perhaps a mild hemiplegia (affecting one side of the body) to severe quadriplegia (very little use of all four limbs) or athetosis.

Some children use electronic aids for written work, and a fairly high percentage have little or no comprehensible speech. However, those with no oral language may have a fair understanding of the concepts discussed in lessons.

Although many of the severe cerebral palsied children cannot actively participate in experiments or recording activities, they do benefit from experiencing purely auditory and visual stimuli.

Their spatial and perceptual problems are often similar to those experienced by spina bifida children.

Thus, when considering the development of a science curriculum in a school for the physically handicapped, it must be remembered that each category of handicap presents different problems.

Orthopaedically Handicapped Pupils

Probably the greatest barrier for these pupils against science in a normal school is the physical environment of steps, stairs, inadequate toilet facilities, wrong bench level and unsuitable desks. Mobility within a classroom is not such a problem, as laboratories in comprehensive schools usually have aisles wide enough for wheelchairs. Such aisles should be kept free of bags, etc., to enable the free passage of handicapped pupils.

If the child has the desire to learn and has the right attitude, then a responsive school can usually cope even with the problems listed above. Special schools are designed to cope with wheelchairs and many of the difficulties can be eliminated. There are usually classroom assistants in special schools who aid normal teaching and help with difficulties when they arise.

If a child has particular manipulative problems, then adaptations to experiments can be made and modifications to apparatus also. Some suggestions are listed in Appendix III.

The book list given on page 213 suggests more extensive analysis of these problems.

Visually Handicapped

This book has assumed that specialised schools for visually handicapped pupils have made their own adaptations to normal science experiments, and certainly many such schools do 'O' level courses in science subjects.

The present text has not concentrated specifically upon the visually impaired pupils' problems, but the booklist refers to some American-based texts published in this area.

Science for visually impaired pupils is at present under investigation in a project at Trent Polytechnic, following lines similar to the science developed for the orthopaedically handicapped.

The SAVI/SELPH project of the University of California would be a most useful reference for all categories of handicapped pupils. SAVI is an abbreviation for Science Activities for Visually Impaired pupils.

Auditory Handicapped

It has been assumed that there is nothing to prevent these pupils doing practical science activities, while appreciating that certain sound experiments are limited unless connected to oscilloscopes, etc. The major problems would appear to be in communication, language development and cognitive growth, but none to debar them from physically doing science experiments.

APPENDIX III

SOME USEFUL AIDS FOR INDEPENDENCE
The experimental section of this book requires only simple apparatus and no specialised or expensive equipment is necessary. However, a number of science schemes listed in Appendix I do use the normal scientific apparatus provided in comprehensive schools. When a handicapped child follows these schemes it is desirable to keep the experimental details as close to the scheme as possible. Some experiments require only minor modifications for use by the handicapped child. For example, often open flames can be replaced by electrical heaters if this is considered a possible risk.

In the course of searching through the many school science schemes and worksheets it has been found that most practical experiments can be adequately modified to allow everyone to do an experiment. In many cases a photocopy of the original worksheet can have some sections or diagrams blotted out and then retyped, and a further photocopy of this modified worksheet can be used by the handicapped child. Most science worksheets have photocopy waiver clauses (these are stated in the teachers' guides).

There are some pupils who are so severely physically handicapped that a more substantial modification is needed and sometimes other classroom aids could be most beneficial, remembering all the time that the aids are only a means of allowing a more effective learning situation.

Some aids are already on the market, similar to those distributed by Nottingham Handicraft. Examples are:

> The Sunflower Selectagrips, a selection of different-sized and shaped handles for cutlery, pens, pencils or any pieces of apparatus.
> The Sunflower handles, designed for easy picking up of

cups but equally adaptable for beakers, jam jars and other pieces of apparatus.

Dycem non-stick surfaces are obvious aids when having to manipulate various items in an experiment.

Stiex Scissors with shaped handles would be invaluable for many pupils with limited hand strengths and mobility.

Gripbit is a plastic material which can be shaped like clay and set solid. Handles can be put onto containers and other objects, so allowing them to be more easily manipulated.

Easiturn Tap Turners would be most useful in the school laboratory and at home.

Hand reachers of various types.

Two potentially useful tools were noticed in a Radio Spares catalogue. These are flexible rods to allow an engineer to pick up screws, etc., which have fallen into an inaccessible part of the inside of a TV set or other mechanical device. One piece of apparatus has a magnet at one end, the other rod has a pair of tweezers. (Radio Spares, PO Box 253, Duddeston Mill Industrial Estate, Duddeston Mill Road, Saltley, Birmingham, B81 BQ).

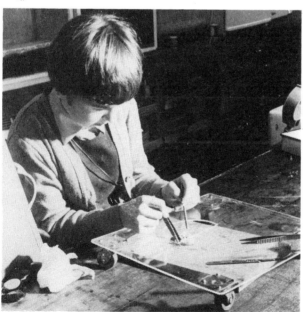

There are many teachers and parents who have made ingenious modifications to apparatus themselves, and some excellent ideas are included in an article in *Special Education* by Gerald Leach of Hephaistos School, Reading.

The types of modification he lists are:

A perspex table top trolley, on Shephard castors to allow movement in any direction; there is also a hole cut in the trolley.

The arms can rest on the trolley and the fine movement of the hands can manipulate apparatus underneath. This was particularly devised to allow a person to dissect a biological specimen through the hole of the trolley. Modifications to this are numerous and a Trent Polytechnic team have tried various shaped cuts. One which looks most widely useful is illustrated here.

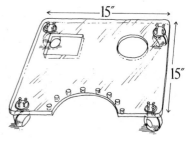

The smaller holes are large enough for a pencil or pen to fit in, allowing for the possibility of drawing shapes. You could possibly think of adaptations more suited to each person's need.

Another Gerald Leach invention is a yoghurt cup sleeve to fit over the arm of a thalidomide child. Any tools needed are pushed through the end of the cup and set into the plaster of paris block inside.

APPENDIX III 211

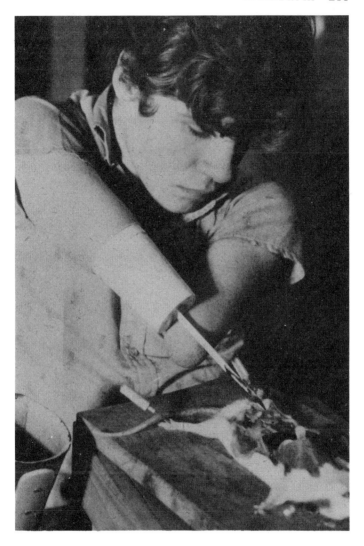

The one mentioned in the article is a pair of tweezers operated by a length of wire to the mouth, which pushes and pulls to close and open the tweezers.

Other inventions listed are bottle fillers, made from washing-up liquid bottles, and a heavy stand for a camping gas bunsen burner and tripod to prevent it being knocked over.

REFERENCES

1. DEAN R. BROWN (1979), 'Science Education and the Handicapped', pages 10–14.
   An article contained in *Sourcebook for Science Education and the Physically Handicapped* published by National Science Teachers Association (Washington USA).
2. B. E. THOMPSON (1976) *Science and Children*, 13 (6), page 5.
3. D. HADARY (1979), *Science Education News*, Fall 1978/Winter 1979, American Association for the Advancement of Science.
4. M. SHAYER and P. ADEY (1981), *Towards a Science of Science Teaching*, Heinemann Educational Books.
5. M. BASSEY (1977) *Nine Hundred Primary School Teachers*, published by Trent Polytechnic.
6. P. BROWN, P. J. HITCHMAN, G. D. YEOMAN (1971), *Schools Council Examination Bulletin Number 21*, Evans/Methuen Educational.
7. Selection of School Science Projects:
   Nottinghamshire Primary Science Project (1979), J. R. N. C. Carter, Trent Polytechnic, Nottingham.
   Look (1981), Cyril Gilbert and Peter Mathews, Addison Wesley.
   5/13 Science (1974), Macdonald Educational and Schools Council.
   Insight into Science (1979), ILEA Curriculum development team, Addison Wesley.
   Science for 70s (1971), A. J. Mee, P. Boyd and D. Ritchie, Heinemann Educational.
   Nuffield Combined Science (1970), Longman/Penguin Books.
   Schools Council Integrated Science Project (1974), William Hall *et al*, Longman/Penguin Books.
   LAMP (1977), Association for Science Education, College Lane, Hatfield, Herts.
   Science at Work (1979), Addison Wesley.
   Open Science (1980), Leon Darrall, Schools Council and Hart-Davies Educational.
   SAVI/SELPH, Centre for Multisensory Learning, Lawrence Hall of Science, Univ. of California, Berkeley, USA.

SOME SCIENTIFIC EQUIPMENT SUPPLIERS

Griffin and George, 285 Ealing Road, Wembley, Middlesex.
Phillip Harris, 63, Ludgate Hill, Birmingham.
Arnolds, Butterley Street, Leeds, Yorkshire.
Nottingham Educational Supplies, 17, Ludlow Hill Road, Melton Road, West Bridgford, Nottingham.

SUGGESTED FURTHER READING

For Teachers and Educators
1 Sourcebook: H. H. Hofman and K. S. Ricker, *Science Education and the physically handicapped*, published by US National Science Teachers Association.
   Some chapter headings include:
   Science for everyone
   Preparing teachers to work with the handicapped
   Resources for teachers/schools
   Science and the auditory handicapped
   Science and the orthopaedically handicapped
   Science and the visually handicapped
   Science program beyond the school
   Careers in Science and related fields
2 SAVI/SELPH (See references in Appendix I)
3 A collection of booklets from the Newcastle upon Tyne Polytechnic. Titles include:
   The use of microelectronics in the education of ESN(s) children, Jim Sandhu.
   The development of teaching material for severely mentally handicapped children.
   Other booklets cover topics such as furniture design and other aspects of educational needs.
4 The other books in the Human Horizons series, published by Souvenir Press.
   M. Cotton, *Out of Doors with Handicapped People*.

5 Hadary, D. E. and Cohen, S., *Laboratory Science and Art for Blind, Deaf and Emotionally Disturbed Children: A mainstream approach*, University Park Press, 1978, Baltimore, Maryland, USA.

6 A number of useful publications from the American Association for the Advancement of Science. Major contributor is Martha R. Redden, AAAS, 1776 Massachusetts Ave, NW, Washington DC, 20036, USA.

Newsletters, Journals, etc

An excellent US newsletter from Dr Ben Thompson:
 Science for the Handicapped Association,
 Department of Education,
 University of Wisconsin,
 Eau Claire, Wisconsin, USA.

Foundation for Science and the Handicapped Inc. Newsletter from:
 Ed. C. Keller,
 236 Grand Street,
 Morgantown,
 West Virginia, 26505, USA.

Occasional Newsletter from:
 School Science Research Project,
 c/o A. V. Jones, Physical Science Department,
 Trent Polytechnic, Nottingham.

D. A. Coleman, 'Technology with the young disabled',
 School Technology, Sept. 1981, p 2 and following.

A. V. Jones, 'Science for the physically handicapped',
 Special Education: Forward Trends, Vol 7, No 3, p 25 f.

G. Leach, 'Making Science more accessible'
 Special Education: Forward Trends, March 1982, pp. 13–15

Sheila Wolfendale, 'The education of handicapped children in the USA – an outline of public law 94–142 and its relevance to policy and practice in Britain,'
 Remedial Education, Vol 17, No 1, 1982, p. 33.

INDEX

Acids 58–71
Aids to learning 26, 209
Aims of science for the handicapped 14–17
Air 31, 143–9
Alkalis 58–71
Ammonia 60
Animals 32
    classification of 47–57
Apparatus 25
    (lists are given before each section)
Astronomy 31

Baked Alaska 100
Balance 173–5
Bicarbonate of soda 60
Blotter fish 52
Boats 120
Bridges: strength of 153–5
    Tacoma Bridge 155
Butterflies 53, 76–7

Cabbage 62
Candles 127–44
Car 32
Cellular structure 31
Cerebral palsy 205
Chemicals, in the home 45–71
Child, development 18–20
    needs 18

Chromatography: 47
    picture 50, 51
Circuits 78–84
Circus method 23
Classification, of animals 46–57
Cold 30
Colour: 46, 62
    changes 68–70
    windows 179
Communication 32, 35
Concepts 20–2, 28
Conductors and insulators 82
Cosmetics 29
Cotton: strength of 164
    woven material 166
Cubes 158
Curtains, net 191

Divers 122–3
Drying out 147

Earth and space 31
Electricity: 28, 31, 72–92
    static 76
    symbols for circuits 94
    writing with 91
Electromagnet 86
Evaluation sheet 33, 34

Fibres, cotton 164
Fire 130

Fire extinguishers 66, 130
Flames 132–8
Flexibility 17
Flowers, colour changes of 70
Foams: 112–19
   bathroom sponge 115
   squashability 118
Food: 28
   chains 30
   webs 30
Forces 31, 168–75
Forgers 49
Friction 31

Geodesic shape 159

Handicap:
   auditory 207
   orthopaedically 206
   visually 206
Hearing 28
Hobbies 29
Home, environment 27
   experiments 42
Hot air balloon 143
Hot and cold 30
House, bricks 156
   of ice 108
   pattern of bricks 156
   straw building 158

Ice:
   expansion of 104–5
   experiments with 95–110
   lifting with string 102
   melting 99, 101, 109
   salt and ice 103
Ideas, USA 14, 202
Illusion 157
Insulators 82

Lamps 139
Leisure 29
Length 30
Levers 31
Lights: 139–41
   polaroid 180
Literacy and numeracy 25
Look science scheme 196

Machines:
   drawing machines 170
   lifting machine 172
   weighing machine 173–5
Magnets 89
Man 31
Materials 30, 151
Melting, of ice 98–110
Metals 30
Methods for teaching science:
   circus method 23
   direct method 23
   problem solving 23
   science interludes 22
   topic work and RBL 24
Microscope 29
Mirrors 142
Motor car 32
Muscular dystrophy 205
Myself 28–9

Network plan 28–9
Nottingham primary science project 196
Numeracy and literacy 25

Ornithology 32

Pantograph 170
Paper, strength of 160
Parents 33, 35, 42–3

Patterns (SCISP) 200
Photography 32
Pictures 71
Plan of curriculum 26
Plants 30, 31, 186
Polaroid light 180
Problem solving 23, 92, 190–2
Projects 87, 149

Question and answer board 83–5

Radio 32
Reactions with acids 58–71
Robots 87

Sahara Desert 104
Salt and ice 102–3
School curriculum 14
Science:
   insights 197
   interludes 22
   processes 22
Science schemes 196–204
   5/13 199
   Insight into science 197
   LAMP 199
   Look 196
   Nottingham project 196
   Nuffield combined science 203
   Open science 202
   SAVI/SELPH 202
   Science at work 201
   Science for 70s 198
   SCISP 200
Secret messages 67
Sellotape 179, 182
Senses 29–30
Shapes:
   cubes 158
   geodesic 159
   pyramid 158
   towers 108, 163
   triangles 162
   twists 188–9
Sight 29
Skills 22
Smell 29
Sound 29
Spina bifida 205
Sporting activities 29
Static electricity 76, 126
Stomach powder 66
Strength of:
   bridges 153–5
   cotton 153
   woven material 164–7
Submarine 124–5

Tacoma Bridge 155
Taste 29
Teachers 42–3, 193
Temperature measurement 145
Time 21
Topic work 20–2, 24, 28
Touch 29
Towers 108, 163
TV adverts 181

Vinegar 60

Washing up liquid 183
Water: 21, 29, 95, 117
   swinging 126
Weight 29
Whammy-diddle 184–5
Windows: 179–80
   net curtains 191
Woven material 166

Zebra 69